Praise for *The Nonlocal Universe*

"Our customary way of viewing consciousness and mind is in the process of radical change—away from the morbid perspective of materialism, to a view grounded in both spirituality and empirical science. In this emerging view, consciousness is considered fundamental, nonlocal and causal, not merely an incidental byproduct of the brain. No individual can consider themselves an educated citizen who is not familiar with these issues, as brilliantly discussed by Dr. Steven Richheimer in The Nonlocal Universe shows how science is not opposed to a spiritual view, but is congruent with it. This book is a gem—comprehensive, authoritative, and written in layman's language. If we are to survive and thrive on this planet, it is crucial that we take these developments with utmost seriousness."

— Larry Dossey, MD, author of *One Mind: How Our Individual Mind Is Part of a Greater Consciousness and Why It Matters.*

"Dr. Richheimer is well versed in science, is a hands-on practitioner of yoga and meditation and is a lucid writer. When you put these qualities together, you get the basis for an important and fascinating account of how modern science supports the views of ancient philosophers and mystics. Richheimer shows how quantum physics describes a world where "all quanta are connected, inseparable parts of a wholeness that underlies our perceived reality." This wholeness is exactly what yogis found in their parallel quest to understand human consciousness and the nature of the universe. If you want to deepen your understanding of physics and at the same time explore what the spiritual worldview offers to humanity, then this is the book for you."

— Dada Vedaprajinanada, Author of *The Wisdom of Tantra*

"Science has uncovered an incredible, almost incomprehensible cosmic dance in which non- locality is the rule, not the exception and the universe is described by wholeness instead of individual parts..." (from the book). The Nonlocal Universe explains how science is not opposed to a spiritual worldview, but is congruent with it. This book is comprehensive, authoritative, and written in layman's language. Highly recommended for anyone curious about our place in the universe."

—Dada Jagadiishananda, yogic monk and meditation teacher with Path of Bliss

"I was a bit daunted by the science going into the book but the author did a good job of making it intelligible to lay person like me. It's great to see how science is coming closer and closer to spirituality and great to have some scientific facts to back it up when you get in those inevitable debates."

— Devashish, author of *In The Land Of The Saints*

"Dr. Richheimer took a very complex idea and made it very readable for me. The attempt to re-link science with spirituality required substantial discussion of physics in the last 100 years. Well done. I have read at least 30 books on the nature of reality and the physical sciences, but very rarely do you get this type of synthesis. While I don't agree with some of the conclusions, the nature of the discussion raised many questions in my mind, as a book of this type should. It's time for some retrospection."

— Michael Pivarnik, Richard E. Fox Charitable Association

"In *The Nonlocal Universe*, Dr. Richheimer has done a masterful job of explaining in easy-to-understand language the remarkable convergence of quantum physics and the ancient philosophy of yoga. The far-reaching implications of this new understanding of reality will have you on the edge of your seat. Fascinating and timely."

— Dada Nabhaniilananda, author of
Close Your Eyes and Open Your Mind

The
Nonlocal Universe

ALSO BY THE AUTHOR

The Unity Principle: The Link between Science and Spirituality

The Nonlocal Universe

Why Science Validates the Spiritual Worldview

Steven L. Richheimer, Ph.D.

InnerWorld Publications
San Germán, Puerto Rico
www.innerworldpublications.com

Copyright 2016 by Steven L. Richheimer

All rights reserved under International
and Pan-American Copyright Conventions

Published in the United States by InnerWorld Publications
P.O. Box 1613, San Germán, Puerto Rico, 00683

Library of Congress Control Number: 2016904824

ISBN: 9781881717492

Cover Design: Rodrigo Adolfo

All rights reserved. This book, or parts thereof, may not be reproduced in any form or by any means, electronic or mechanical, including photocopying, recording, or by any information storage or retrieval system, without permission of the publisher except for brief quotations.

Dedication

This book is dedicated to my loving guru, Shrii Shrii Anandamurtiji), whom we affectionately call Baba. He has been my spiritual guide and inspiration for nearly five decades. He was a personality incomparable to any other and undoubtedly the greatest spiritual teacher of the twentieth century. He has been an inspiration to me and to millions of his followers.

Acknowledgments

I would like to thank my wife and spiritual companion, Jeanne P. Richheimer, for her help in editing this book and for helping with the illustrations. I am also deeply indebted to Devashish Donald Acosta for his help in editing and laying out the final manuscript for this publication.

Contents

Preface	1
Introduction	3
PART I **NONLOCALITY OF THE** **PHYSICAL WORLD**	9
1. How it all Began	11
2. The Nonlocality of Quantum Mechanics	17
3. More Quantum Weirdness	22
4. Nonlocality of Time and Space	31
PART II **NONLOCAL MIND**	47
5. Why Mind is not Brain	49
6. Mystical Experiences	59
7. Remembering Past Lives	64
8. Out-of-body and Near-death Experiences	70
9. Extrasensory Perception	78
PART III **SPIRITUAL IDEOLOGY**	95
10. The Cycle of Creation	97

11. Mind, Body, and Death	112
12. Explaining Scientific Anomalies	130
13. Science and Metaphysics	142
14. Karma, Happiness, and Suffering	154
15. The Quest for Limitlessness	160
16. The Problem with Ego	168

PART IV
CAN THE SPIRITUAL WORLDVIEW
SAVE HUMAN SOCIETY? 177

17. Nonlocality and the Existence of God	179
18. The Problem of the Day	189
19. The Spiritual Worldview to the Rescue	195
Notes	202
Index	210

Preface

Nonlocal:

Designating an operation or property of a topological space that is not applicable only to a specified neighborhood in the space; specifically in quantum theory, relating to the phenomenon whereby properties of isolated systems which are not causally related classically or relativistically may depend on each other.

My purpose in writing this book is to prove that recent discoveries in the physical sciences—particularly quantum mechanics and relativity theory, indicate that nonlocality is a basic feature of physical reality. There is also conclusive evidence that mind is nonlocal. This leads to the conclusion that the only consistent explanation for reality is a spiritual one, which begins with Consciousness as the first cause.

The fact that both space and time are now understood to be nonlocal phenomena may be new to many readers of this book. However, to a scientist schooled in quantum mechanics or Einstein's theories of relativity it will not be new or surprising, but it may still seem weird, mysterious, and inexplicable.

To the common person in the West, the phenomenon of nonlocality challenges the "common sense" understanding of reality. It undermines the predominant worldview, which is materialistic. No less than a paradigm shift in thinking is needed before humankind can begin to understand a new worldview that infers that everything in the universe is interconnected and ultimately one. Essentially this is the worldview of spiritual ideology. It may come as a surprise to many readers that the origin of this worldview dates back approximately seven thousand years to the great philosopher and spiritual master Sadashiva (Shiva). In contrast to the materialistic Western worldview of reality, which is strongly dualistic, the Eastern or spiritual worldview can be termed monistic or monistic idealism. That is, everything can be reduced to one, and all the differences observed in everyday life are a relative reality. The Ultimate Reality is the One or Cosmic Entity and relative reality is ultimately an illusion.

Introduction

THE SEPARATION OF SCIENCE and spirituality can be traced back to the seventeenth-century philosopher, René Descartes. He said that the body works like a machine and is therefore material, while the mind (or soul) is nonmaterial and does not follow the laws of nature. Hence, according to Descartes, mind fell into the domain of religion (or spirituality) and matter fell into the domain of science.

The discoveries of Newton and other eighteenth- and nineteenth-century scientists served to reinforce this view of science and left any discussion of spirituality to religion. The mechanistic Newtonian physics advanced our understanding of the motions of celestial bodies and physical objects, while at the same time, Darwin provided a reasonable explanation for the evolution of complex living organisms.

According to this new model of classical physics, objects, no matter what size or mass, were governed only by direct interaction or local force fields and were limited by the speed of light (local matter-energy interactions). Furthermore, if initial conditions were known and the laws of physics applied, then all phenomena could theoretically be predicted exactly. Finally, mind and consciousness originated from matter and therefore could not exist without a brain and nervous system. Mind and consciousness as epiphenomena of matter could have no direct effect upon physical objects and were merely products of brain activity (epiphenomenalism). This worldview assumed upward causation. That is, it starts with elementary particles that form atoms, which form molecules, which form living cells, which compose the brain, which creates mind, and finally mind creates consciousness.

Materialism even had an explanation for how life originated from random chemical permutations and how more complex living organisms and structures resulted from random mutations and selection of the fittest (material Darwinism).

Nineteenth-century scientists were so successful in explaining the many mysteries of the universe as the workings of a finely tuned mechanistic system, much like a clock, that they believed that they were on the verge of explaining everything in terms of matter, energy, and physical forces. There was no longer a need to draw upon a supernatural being or God to explain the origin or basis of reality.

The materialistic approach of science has been very successful in explaining natural phenomena; and technologies based on the discoveries of science have transformed our world for the better. The accomplishments of science and technology for improving socioeconomic conditions of humankind should not be underestimated. However, these successes have also persuaded a majority of people that the materialistic worldview of science is a true and accurate description of reality.

In Part I of this book, we look at why the "new physics" describes reality as nonlocal. This story began with the discovery of an anomaly in 1887 by two American scientists, Albert Michelson and Edward Morley. They performed a simple experiment that demonstrated that light did not behave the way scientists expected. It was well known that light displayed wave-like properties. For example, light waves diffracted and interfered just like other waves. Physicists naturally assumed that just like water or sound waves, light needed a supporting substance or medium through which it could move. This medium was termed the "luminiferous aether," and it could transmit the wave motions of light—even through the vacuum of space.

The crucial experiment conducted by Michelson and Morley used an interferometer that could detect very slight differences in the speed of light that would be expected if their instrument was directed into the flow of the aether or against it. It was assumed that light from a distant star would behave like any other moving object and its velocity would add to that of Earth in one direction and be less in the opposite direction as Earth moved through the stationary aether. However, this did not occur. Instead, the speed of light was a constant no matter which direction their instrument was pointed. The Michelson-Morley experiment was the first nail in the coffin of the then-prevalent aether theory, and as other experiments confirmed their results a new line of research was begun that eventually led to Einstein's special theory of relativity.

A second experimental observation that did not fit the mechanistic model of the universe was the emission spectrum of atomic hydrogen. Hydrogen is the simplest and most prevalent element in the universe. It consists of a single proton and electron. When hydrogen is heated, its electron can become "excited" or move further from the proton that lies at the nucleus of the atom. Upon cooling, the electron will return to a "more comfortable" lower-energy level and release a photon of light in the process.

Classical mechanics would suggest that such transitions should emit photons of various wavelengths resulting in a continuous spectrum of light. In actuality only a series of spectral lines were observed, suggesting that the electron can only make specific transitions between different energy levels in the atom. The discovery that electrons could only occupy specific energy levels or orbits and would seemingly jump between these levels without passing through intermediate levels (so-called "quantum leap") was crucial for the development of quantum mechanics.

Both relativity theory and quantum mechanics have been hugely successful in explaining a multitude of physical processes and phenomena both on very large and very small scales. We now know that Newton's vision of a "clockwork" universe in which all of space can be mapped out in three dimensions with all clocks ticking at the same rate is false. Space and time are understood to follow relativistic mechanics in which both space and time form a fully integrated continuum, known as space-time.

Moreover, today we know that tiny pieces of matter and energy do not behave classically. Energy consists of discrete packets called "quanta." Sometimes these quanta behave like waves and sometimes like particles. Matter and energy are interconvertible and matter is now considered a condensed form of energy.

These discoveries of modern science describe a reality that is far more complex and rich then any imagined by earlier scientists. Virtually all the assumptions underpinning the nineteenth-century materialistic worldview have been negated by scientific discoveries of the twentieth and twenty-first centuries. Adherents of materialism (aka material realism, physicalism, reductionism, and scientism) had to reinvent the doctrine in order for it to fit the observations that contradicted the earlier version. As we will see, this "new" version of materialism is more in tune with today's scientific observations, making it a viable worldview that is accepted by many in the scientific community. However, the "new materialism" has had to propose various untestable hypotheses and has failed to address nonlocality and other psychophysical and psycho-spiritual phenomena and observations.

In Part II of the book, we explore why mind cannot be equated with brain (matter). This is one of the fundamental doctrines of materialism, but the hypothesis that mind and consciousness are epiphenomena of brain does not fit the facts. Studies of numerous psychophysical phenomena and psychic phenomena such as out-of-body and near-death experiences, reincarnation, mystical experiences, and extrasensory perception (ESP) offer incontrovertible proof that mind is nonlocal. Moreover, modern science has now proven that the building blocks of space, time, matter, and energy are not localized. Why would mind and consciousness be local? These are subtler than matter.

Countless experiments have shown that quantum events are dependent on their observation. The simple act of observing or measuring a quantum system will change the system. At best, one could say that the objectivity of quantum mechanics is weak, in that the outcome of the observation of a quantum event is not dependent on who makes the observation.[1]

These and other scientific observations directly contradict the principle doctrine of materialism—that everything can be reduced to the interplay of matter and energy, and that there is no need to incorporate the "superstitions" of religion, spirituality, or consciousness into any discussion of reality. It seems most scientists are reluctant to bury the corpse of materialism and allow consciousness to take its rightful place in science. Perhaps the reason for this is that to do so would be to open the scientific realm to metaphysics and allow an idealist revolution to sweep away the outdated doctrine of materialism. The problem is that the materialistic worldview is a dualistic theory that tends to degrade the human mind by saying only the physical world is real. Most, if not all, of society's problems can be traced directly or indirectly to this philosophical model of reality. Now that experimental observations contradict the materialistic model of reality there is a need to reject the old model and embrace a new model that accepts the existence of a vast, unseen transcendental reality, which places humankind at the pinnacle of creation—a realm founded in Consciousness with infinite possibilities and indescribable bliss.

In Part III of this book, we discuss this new model of reality and why it explains all the "anomalies" that the reductionist worldview cannot. It can be called the "Eastern worldview" or "spiritual worldview." It may also be termed "spiritual ideology." This worldview is distinguished by monism or monistic idealism—the idea that there is only one infinite, absolute, Ultimate Reality. To many in the West, the idea that the created universe is one undivided whole is a foreign concept that goes against everyday experience. It implies that discreetness, differentiation, individualism, etc. are illusory. If the ultimate reality is the One, everything else falls under the category of relative reality—important in our day-to-day lives, but ultimately an illusion.

In this worldview, Cosmic Consciousness (Consciousness) takes center stage. Consciousness is considered the "ground substance" of creation. It is Consciousness that is transformed into the material world. Creation is believed to begin with unqualified awareness or Consciousness, which under the influence of a qualifying principle is gradually transformed into the subtle feelings of "I am," "I do," and "objective I." Further qualification of this "objective I" ("mind-stuff") creates matter and energy, which form the physical universe. Living organisms evolve in the second stage of this cosmic cycle, which constitutes a countermovement from crude to subtle.

The spiritual worldview is central to the religious traditions of the East. These include Vedantism, Buddhism, Tantra, yoga, Taoism, and Sufism. One common theme of these philosophies is that creation begins with Consciousness and one's individual existence continues until it is merged or lost in this unqualified sea of pure Consciousness (i.e. God). In other words, the creation is cyclical. It begins and ends in the unqualified Consciousness.

Unlike the philosophy of dualism, which teaches that the universe consists of, or is explicable as, two or more fundamental entities, such as matter and mind, living and nonliving, God and his creation, etc., the monism of the spiritual worldview purports that the universe is a Singularity. Everything is connected, and all things are simply manifestations of the Cosmic Entity (Brahma). This is precisely the model of reality as revealed by quantum mechanics. The wave function of quantum mechanics is not one of separate parts but one in which all possibilities coexist in a state of wholeness, everything being interconnected and interdependent.

We should be very clear about one thing, however—spirituality is not religion. There have been many well-documented failings of organized religion.[2] Spirituality, on the other hand, remains unscathed. It can be described as the foundation for all the world's great religions. The great religious teachers of the past all appear to have experienced a spiritual rebirth. Such a "mystical experience" of the wholeness or oneness of creation has been described and called various names. These include cosmic consciousness, nirvana, enlightenment, liberation, self-realization, samadhi, etc. It appears that the degree to which the religions of today deviate from spiritual ideology is simply a measure of how their followers have corrupted the original spiritual message of their founder. One may discount the scientific validity of such experiences, but there is no denying the fact that people who have experienced a mystical connection or union with God have been truly changed by these experiences, as have many of their followers.

In the last part of this book, we discuss why nothing less than a paradigm shift of human consciousness is needed if human society is to progress and survive in the future.

PART I

NONLOCALITY OF THE PHYSICAL WORLD

1
How it all Began

THERE IS STRONG AND convincing scientific evidence that both time and space began in a gigantic explosion that occurred some 13.8 billion years ago in what is commonly called the Big Bang. Beginning from a dimensionless point that was incredibly hot and dense, the universe initially expanded much faster than the speed of light while at the same time the first subatomic particles were created that later cooled and formed the simplest elements consisting of hydrogen, helium, and lithium.

Major evidence supporting this theory comes from Edwin Hubble's discovery that distant galaxies are moving away from Earth and from each other. This rate is proportional to the distance separating them from each other (Hubble's law). Hence, if a galaxy is one hundred light years distant, its velocity might be V. However, for a galaxy that is two hundred light years distant it is found to be receding from Earth at 2V.

Big Bang theory offers a comprehensive explanation for a broad range of observed phenomena, including the abundance of light elements, the cosmic microwave background radiation, the large-scale structure of the universe, and Hubble's law.

The expansion of the universe cannot be due to the mechanical results of the Big Bang but must result from the expansion of space itself. If this were not so, then distant galaxies would be moving away from us at a speed greater than light, which is not possible. Another consequence of this type of expansion is that the universe, which began as a point, has no center. No matter where you stand in the universe, everything is observed to be receding from you at the same relative rate depending on its distance from you. Hence, the center of the universe can be considered to be anywhere or everywhere, including where you are now standing.

The discoveries and theories of Einstein have shown that time and space are complementary aspects of the same thing. Hence, time and space are not independent entities but are fully entwined with one another in a four-dimensional continuum called space-time. This means that space is free to convert to time and time into space. Although the linear concept of time plays an important role in the physical sciences such as geology, astronomy, physics, and biology, linear time is a classical concept. In both relativistic and quantum physics, time is not linear since it is inexorably tied to space, and an effect may precede the cause.

Cosmologists assert that space and time did not exist prior to the Big Bang; therefore, it is meaningless to try to hypothesize about what came before it. However, cosmologists are free to speculate on why the Big Bang occurred and what came immediately after it.

For example, cosmologists now postulate that there was a time in the past when all the forces and quanta of the emergent universe were unified. They call this period the "Planck epoch" and it occurred only at the very beginning of the universe, some 10^{-42} seconds following the initiation of the Big Bang.

Problems with the model

The expanding universe model requires that space expand at a certain rate since the time of the Big Bang. However, this model has several problems built into it. One problem is the size of the universe. If the universe began as a point and first expanded at the speed of light then after one year its size would be one light year (the distance light travels in one year). After approximately 14 billion years, its size should be 14 billion light years. However, the observed size of the universe is some 46 billion light years. This would not be possible unless the initial expansion was much faster than the speed of light.

Secondly, the universe is known to be extremely homogeneous, but the mixing that is required for homogeneity could not take place if separate sections of the early universe expanded at or near the speed of light. There would simply be no way they could have had contact with one another. To account for homogeneity one must assume that all quanta are entangled and/or the initial expansion of space took place at a speed greater than that of light.

In addition, physics has no explanation for why more matter particles emerged from the Big Bang than antimatter particles. All the theories indicate that in a universe born in a tremendous explosion such as the Big Bang, there should be equal amounts of matter and antimatter. If this were the case, then the colliding pairs of matter and antimatter particles would annihilate each other leaving nothing but energy and empty space. Since there was obviously an excess of matter particles, physicists have had to propose untestable theories to explain this.

Another difficulty with the model is that of a flat universe—one that neither expands rapidly nor contracts rapidly. If the expansion of space is even a tiny bit slower than the counteracting force of gravity then the universe is said to be "closed," gravity would prevail, and the universe would begin to contract leading to a "Big Crunch." The rate of contraction would accelerate rapidly leading to a universe that collapses back to a point in only a few million years. On the other hand, if the rate of expansion of space were slightly larger than the counterbalancing pull of gravity, then the result is an "open universe." In this case, the rate of expansion would accelerate as mass and thus gravity became diluted. After a few million years, everything in the universe would be so far apart that stars and galaxies could never form. There is a very fine line between these two models, but the longevity that results from a flat universe is crucial, for it provides the time needed for stars, galaxies, planets, and eventually life to evolve.

Another problem with the Big Bang model is the energy of empty space. This energy is called the "cosmological constant" or "lambda," and it could be either positive (attractive) or negative (repulsive). It represents the gravitational force of space itself. Originally, scientists believed in a static universe that neither expanded nor contracted. This factor was needed in order that the universe not collapse under the force of gravity but remained of fixed size. However, Hubble's discovery of cosmic expansion disproved the static universe model, and at first there was no need for a fudge factor such as lambda to halt an expanding universe. However, scientists now find that more distant stars are receding from us at an accelerating rate, not at all the rate that would be predicted if space was expanding regularly against the force of gravity. In this case, there should be a slight deceleration in the expansion. Lambda, the so-called cosmological constant, is again needed; and as a result empty space must have more energy than first believed.

This energy of the void is called "dark energy" and it is believed to compose some 75 percent of the mass-energy of the cosmos. Scientists still do not understand what it is, but it must have some weird properties since it is both gravitationally repulsive and not diluted by the expansion of space. Apparently, dark energy acts like a rubber band that increases in tension as space expands and this increase in tensile force counteracts the effects of dilution.

In addition to the mysterious dark energy, cosmologists also recognize the need for another mysterious substance that has been labeled "dark matter." Calculations show that there is not nearly enough ordinary or observed matter in galaxies to hold them together. There must be a relatively massive amount of unseen matter in galaxies to account for their formation and stability. Estimates place the amount of dark matter at 20 percent of the mass-energy of the universe. Currently scientists have no clear idea what this unidentified matter is. Since dark energy and dark matter constitute some 95 percent of the mass-energy of the cosmos, the normal matter that is observed in all the stars, galaxies, black holes, etc. accounts for only 5 percent of the total. Hence, scientists still have a long way to go in their understanding of the cosmos.

Since the total curvature of the universe appears to be very close to zero or flat, this means that the total mass-energy of the universe, which includes ordinary matter, dark matter, and vacuum energy (dark energy), has to add up to the critical density required for a flat universe. The likelihood of this scenario occurring by chance is extremely small, leading to the problem of how the universe became "fine-tuned."

To account for the various anomalies of the Big Bang model of cosmogenesis, scientists came up with the idea of "cosmic inflation." According to this theory, the universe initially expanded very rapidly—much faster than the speed of light. Since it was space that expanded, this does not contradict the theory of relativity, since the speed of light was not affected. At the end of inflation, the universe is thought to have decayed into normal matter and energy (radiation), and the temporary hyperfast expansion gave way to the normal expansion of the universe observed today.

The inflation theory solves most of the problems of the Big Bang theory of cosmogenesis. Theoretically, all the parts of the universe began fully connected in the inflation phase and like completely mixed milk and coffee everything was homogeneous. Hence, the theory of inflation explains why the temperatures and curvatures of different regions of space are so constant. The theory also allowed physicists to predict the minute differences in temperature of different regions of space in the primordial universe from quantum fluctuations during the inflationary era, and these predictions have been confirmed by observations of the remnants of the Big Bang—the cosmic microwave background radiation. Because of the very rapid growth of space during the inflation phase, the universe expanded far beyond the size predicted if it were limited to the speed of light.

Inflation also provides a mechanism whereby the universe could initially emerge in a flat state following inflation. As we have seen, flatness is both unstable and improbable. Even a slight excess of matter-energy as compared to the power of expanding space (lambda) results in a closed universe that very rapidly contracts to a Big Crunch. On the other hand, if expansion is slightly greater than the gravitational forces, then an open universe will result that leads to emptiness that is unable to support life. The theory is that when ordinary matter finally condensed following inflation, the tremendous tension in lambda that was built up during the superfast expansion suppressed the curvature of the universe, after which it reverted to its normal strength that we witness today. The theory depends only on the idea that lambda is not diluted by expansion and therefore can overcome the forces that lead to curvature of the universe. Hence, inflation theory explains why the curvature of the universe would be suppressed and might help explain the required tuning that was necessary for a flat universe to emerge following this phase of superfast expansion.

However, inflation theory simply substitutes one set of improbabilities for another. The temporary lambda of inflation becomes a normal lambda after inflation, and the new value for lambda still needed to be "just right" so that our universe, which emerged from inflation, had the perfect balance of gravitational energy and mass-energy so that it neither hyperexpanded or hypercrunched. In the end, the fine-tuning problem is not resolved, since a precise value for normal lambda and the total mass-energy of the universe is still needed in order to account for our long-lived universe. In fact, Stephen Hawking calculated that the odds for this occurring by chance are only one in a million trillion.[1]

In order to explain this enigma of fine-tuning, cosmologists have proposed multiple universes or "multiverse" theory. Perhaps our universe has the perfect balance of factors because an almost infinite number of universes were formed after various regions condensed or emerged from inflation—much like the production of a long sheet of bubble wrap. All these universes are separate; they do not interact with one another and each collapsed with slightly different values for the cosmological constants, mass-energy, and physical laws that govern it.

Along with the multiverse theory to explain why our universe seems to be fine-tuned for conscious life, one needs to apply the so-called "anthropic principle." This principle states that we could only exist in a universe that was finely tuned for our conscious existence. Thus, while the probability might be extremely small that conditions conducive for life exist in one of these universes, the almost infinite number of possible universes insures that there must be a finite number of life-supporting universes.

2
The Nonlocality of Quantum Mechanics

BY THE EARLY TWENTIETH century, the "classical physics" of Newton seemed to explain almost all phenomena of the physical universe. Only a few "anomalies" remained unexplained. As was mentioned earlier, one of these was the constant speed of light as evidenced by the Michelson-Morley experiment. Another was the emission spectrum of hydrogen with its distinct lines instead of a continuous spectrum of light, and a third was black-body radiation. In order to explain the observed emission of energy from a small hole in a large cavity (black body), Max Planck had to assume that the energy of the oscillators in the cavity was quantized, i.e., oscillations occurred in integer multiples of a constant quantity. Later Einstein built on this idea and proposed in 1905 the quantization of electromagnetic radiation to explain the photoelectric effect. Scientists at the time thought that these relatively minor inconsistencies of classical physics would be easily resolved by refinements in classical theory. However, it turns out they could not have been more wrong.

In the early part of the twentieth century, Neils Bohr and his colleagues in Denmark were the first to tackle the problems of the hydrogen emission spectrum, black-body radiation, and the photoelectric effect—all of which were known to be quantized. The theory they developed to explain these phenomena is called "quantum mechanics." The theory held, for example, that the electron in a hydrogen atom can only exist in certain energy levels or "orbitals," and when the electron transitions between levels only specific quanta of light (photons) having specific wavelengths can be emitted. A similar conclusion was arrived at by Einstein to describe what was happening with black bodies and the photoelectric effect—discrete quantized packets of energetic light particles (photons) with a certain threshold frequency are required to emit electrons.

As quantum theory developed, it became clear that quantum particles had to be nonlocal in nature. That is, their position and momentum (speed) cannot be known with certainty; at best, one can only assign a probability that the particle will be found at a particular location. The theory also predicted that two or more paired quantum particles do not behave as separate entities. The rules governing the new theory required that, for example, when the spin of one of two correlated quantum particles is measured it affects its partner, even if the two objects are on opposite sides of the galaxy, and the change occurs instantaneously. The particles are said to be entangled.

One cannot overemphasize the challenge that this singular prediction presented to the prevailing scientific attitude of the time. Not only did it challenge everything scientists believed about the nature of physical reality, i.e. classical scientific epistemology, but it also went against the prevailing idea that for every physical phenomenon there must be a corresponding physical theory to explain it. This prediction of quantum nonlocality was so weird, so crazy, so impossible to understand, that most physicists at the time, including Einstein, believed that it demonstrated an incompleteness or error in quantum theory.

Einstein and his colleagues believed that this prediction of quantum mechanics represented its Achilles heel. They argued that according to the prediction, quantum particles separated in space would have to interact with one another instantaneously via local signals at a speed greater than that of light. Einstein called this "spooky action at a distance." Einstein believed that signals between particles could never exceed the speed of light, and if quantum mechanics predicted faster-than-light communication between particles, then there had to be a serious flaw with the theory.

Einstein argued that if, as experiments at the time suggested, observation of a particle with a spin in the up direction insured that its twin had a spin in the down direction then this meant that their spins were simply fixed from the time they were born. Just as if you had a pair of gloves, placed them in separate boxes, and sent one to Berlin where it was opened and found to contain a right-handed glove, the researcher there would know that the other box still in Boston contained a left-handed glove. No mystery, no weirdness, local Einsteinian epistemology is safe! Bohr on the other hand pointed out that one could not possibly know the quantum state of the particle until it was measured or observed. It was free to fluctuate randomly between the up and down spin states and so too was its partner. The real issue here was separability. The material realism of Einstein considered all objects as separate entities, while Bohr's new physics of quantum mechanics required that quantum particles be nonlocal and hence connected through space-time. Einstein and Bohr, along with other scientists, had a running feud about the nonlocality of the quantum realm that lasted nearly twenty-five years but was finally resolved by experiments performed after the death of both men.

Proof of quantum nonlocality

In 1964, physicist John Bell proposed an experiment that would conclusively prove which version of reality was correct. Like many physicists at the time, Bell both hoped and believed that testing his theorem would prove that physical reality would hold up to physical theory. In other words, that particles can only be influenced by local signals or fields. A test of Bell's theorem would show that either quantum mechanics was consistent with reality or it was incomplete and flawed. It would also prove which of two fundamentally different views of reality were correct. However, it took another twenty years before the experiments could be conducted that would test Bell's theorem and conclusively prove that nonlocality was a fact of nature.

The definitive experiment testing Bell's theorem was conducted by Alain Aspect and his colleagues at the University of Paris-South in 1981. Aspect set up an apparatus that emitted paired photons. That is, the two photons were emitted simultaneously from a calcium atom with the same quantum characteristics except for their polarization. These were in a singlet state, or correlated quantum particles. According to quantum mechanics, such particles are intimately connected, and it is not necessary for any signal or field to pass between the particles for one particle to know what the other is doing. Hence, if the polarization of one is changed or measured, quantum theory requires that the polarization of the other particle must change simultaneously in such a way that the polarization of the two particles always remains opposite—i.e. one vertical, the other horizontal.

Polarization is simply the angle at which the electrical vibration occurs relative to a plane for electromagnetic energy such as light. Most people are familiar with polarized sunglasses that block horizontally reflected sunlight, reducing glare. The detectors that Aspect used were similar to such glasses, except that they could be turned to any angle at will. The uncertainty principle, which applies to quantum objects, requires that the polarizations of the two correlated photons are indeterminate until one is measured, but when this occurs then the other photon must have the opposite polarization.

As predicted by quantum theory the two photons were always found to have opposite polarizations. The test of Bell's theorem eliminated the possibility that the particles were created with opposite polarizations that never changed, which had been Einstein's argument. These observations alone did not prove nonlocality, since it was possible that the two photons could communicate via local signals at the speed of light. However, Aspect added an electronic polarization switch that had the effect of changing the polarization setting of one of the detectors every ten-billionth of a second—shorter than the time needed for light to travel between the two detectors. Nonetheless, the change of the polarization setting in one detector influenced the outcome of the measurement in the second detector, showing that the two photons were indeed connected through space as predicted by quantum theory. There was simply no time for the information about the change in one detector to be transmitted to the other paired photon through local signals.

In 1997, Nicolas Gisin of the University of Geneva and his colleagues sent linked pairs of photons along fiber-optic cables in opposite directions to villages north and south of Geneva. While in the Aspect experiment the spatial separation between the particles was only thirteen meters, Gisin and colleagues had their detectors eleven kilometers apart. According to classical physics, there would be no way these photons could communicate with each other. However, when the independent polarization of one photon was compared to its twin their polarizations were always opposite. As in the Aspect experiment, the photons were entangled. They shared a common origin and properties and remained in instantaneous touch with each other, no matter how far apart they were. Quantum theory's controversial prediction of nonlocality was confirmed. There could be no other explanation for these observations. Classical epistemology was shown to be false. Nonlocality was a feature of reality and a new epistemology based on this fact would have to take its rightful place in man's understanding of reality.

Since cosmology suggests that all quanta were unified right after the start of the Big Bang, and experiments verify that quanta that are born together remain unified through all time and space, it follows that all quanta in the universe must be considered entangled. Hence, the simple explanation for nonlocality is that the universe consists of an undivided whole in which there are actually no discrete or separate parts. The illusion of discreteness only manifests when there is observation of an event with either instruments or awareness. This is not to say that the things humans observe do not have physical reality—just that the sum of the parts can never approximate the whole. The parts are constituents of a relative or physical reality, while the whole constitutes "Ultimate Reality." Science can only deal with the relative reality of the parts; it can never begin to measure the whole, since any attempt to measure the whole using its parts is doomed to fail. However, this does not mean that all attempts to understand the whole should be abandoned, nor should science refrain from the study of the physical universe, mind, and consciousness. The inferences generated by such studies are very important. They indicate that the cosmos consists of an indivisible whole. There does not appear to be any other reasonable explanation.

Although the proof of quantum nonlocality has enormous consequences for how reality must be viewed, there has been relatively little attention paid to it. Even scientists are confused and somewhat amazed by the implications that nonlocality in the quantum realm might have on how we view the nature of reality. Thus, it is no surprise that most of the public has little, if any, understanding of the implications of this discovery.

3
More Quantum Weirdness

> The elementary particles are certainly not eternal and indestructible units of matter; they can actually be transformed into each other. The world thus appears as a complicated tissue of events, in which connections of different kinds alternate or overlap or combine and thereby determine the texture of the whole.[1]
> —Werner Heisenberg

QUANTUM PHYSICS STRONGLY SUGGESTS that everything in the universe is connected. However, there are other aspects of the theory that fall under the category of weird, yet support the idea that matter-energy interactions at this tiny level are consistent with a universe that is a singularity.

Any discussion of quantum mechanics would be incomplete without a discussion of two of its principle features: indeterminacy and complementarity. Indeterminacy simply means uncertainty. It is one of the cornerstones of quantum mechanics. It was first described by Werner Heisenberg in his famous uncertainty principle, which states that the uncertainties in measuring the position and the momentum of a quantum particle are always equal to or greater than a constant (Planck's constant). That is, the greater the precision with which the position of a quantum particle such as an electron is measured, the less precisely is its speed known, and vice versa. Inherent in this principle is the phenomenon that the very act of measuring the position or speed of the particle affects it, and this causes uncertainty in its measurement. We can ignore the fuzziness, which is characteristic of the quantum realm in everyday life, because it is so small. Because of this, the classical descriptions of nineteenth-century physics normally work well on the macro scale and only fail on the micro scale.

Complementarity in the quantum realm describes the entangled and ambivalent qualities of quantum particles or phenomena. For example, the dual aspects of light as both wave and particle are complementary. Neither description for light works under all experimental conditions. Depending on how an experiment is designed, light will sometimes behave like a wave, while under different experimental conditions it will behave like a particle. A complete description of light requires that both of these mutually exclusive constructs be considered, and knowledge of the situation is necessarily limited because both descriptions cannot be simultaneously measured or described precisely (uncertainty principle).

The existence of complementary aspects for describing a physical construct indicates that the true reality of the construct is to be found at a deeper level than that of the two complementary descriptions. Examples of complementarity in the physical realm are wave-particle, space-time, position-momentum, and matter-energy. Today complementarity is the logical framework for comprehending and acquiring scientific knowledge in the physical realm. The concept of complementarity can also be applied to other fields besides quantum mechanics when there is no single or simple way to describe reality from a limited perspective. Examples of this type of complementarity are particular-contingent, cause-effect, observed-observer, thought-action, physical-psychical, yin-yang, male-female, and part-whole.

One of the weirder aspects of quantum mechanics is that on the quantum level something can simultaneously exist and not exist. For example, if a subatomic particle is capable of existing in several different states, the uncertainty principle allows it to exist in all possible states at the same time. When the particle is measured or observed by physical means then its state is no longer uncertain. The act of measurement instantly forces it into just one state. This is called the "collapse of the wave function." As we have seen with particles that are "paired," the observation of one particle causes its wave function to collapse into one state while simultaneously the wave function of its twin collapses into an opposite state, even if the particles are separated by great distance.

Physicists also observe quantum particles emerging from the "vacuum" of space-time only to disappear again. High-energy scattering experiments of the past few decades have demonstrated the dynamic and ever-changing nature of the particle world in the most striking way. Matter has appeared in these experiments as completely mutable. All particles can be transmuted into other particles; they can be created from energy and can vanish into energy. In this world, classical concepts like "elementary particle," "material substance," or "isolated object" have lost their meaning; the whole universe appears as a dynamic web of inseparable energy patterns.

Experiments demonstrating temporal and spatial nonlocality

The study of quantum physics leads inexorably to the conclusion that the quantum world exists in both temporal and spatial nonlocality. What is meant by temporal nonlocality? Simply put, it implies that the past, present and future coexist in the now and that time does not flow linearly from the past to the present to the future. For example, in some cases effects are found to precede the cause.

Spatial nonlocality means that quantum particles cannot be placed with certainty at any given set of coordinates. They are described mathematically by a wave function, which provides only probabilities of where they will be found when observed or measured; and this wave nature of quanta is not localized in either time or space, which means there is always a finite probability that they can be found anywhere in the universe.

One of the classic experiments of modern quantum physics that reveals the nonlocality of space and time is the dual-slit experiment. In this experiment, light (photons) or electrons are passed through two slits that are very close together. Both electrons and photons have wave and particle characteristics depending on how they are observed. Typically, the wave-particles passing through the slits interact as waves to form an interference pattern (alternating dark and light lines) as the waves either cancel or reinforce one another. This is exactly what is expected if light behaves like a wave.

However, light can also act like a flow of particles; for example, if its intensity is lowered to the point that only one photon passes through one of the slits of the apparatus at a time. In this case, the interference pattern might be expected to disappear since it is created by the interaction of wave fronts from the two slits, and the individual photons arriving at the slits at different times should not be able to interact. However, the pattern persists even though the individual photons cannot split, each half going through a different slit and then recombining before they hit the detector. The interference pattern may be observed by exposing a photographic plate for many minutes or summing the response from an electron detector. The existence of the interference pattern under these conditions suggests that the so-called "well-defined" particles are behaving like a wave and individual particles are interacting with particles that passed through the slit at different times. That is, they exhibit both spatial and temporal nonlocality.

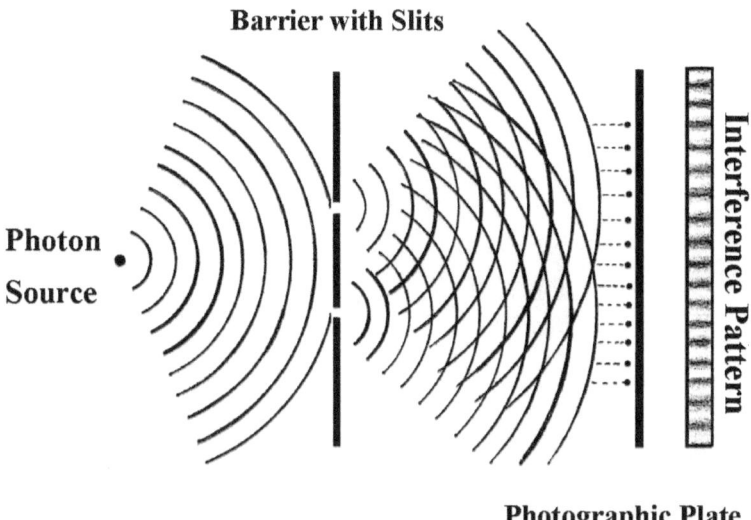

Dual-slit experiment. Waves from the photon (light) source pass through two closely spaced slits causing the wave fronts to interfere with each other producing alternating dark and light lines.

In another experiment, an apparatus is used that has a very thin coating of silver that creates a 50 percent chance that a photon will pass through and a 50 percent chance it will be reflected.

If the photon passes directly through the half-silvered mirror, it takes a direct path through the slits onto the photographic plate, while if it is reflected it must take a longer route and arrive at the photographic plate later. It should be an either/or situation and there is no reason to expect the photon to interfere with itself and create an interference pattern. However, development of the photographic plate reveals an interference pattern, which indicates that the photon is somehow enmeshed with itself across time. This is a classic example of temporal nonlocality at the quantum level.

26   THE NONLOCAL UNIVERSE

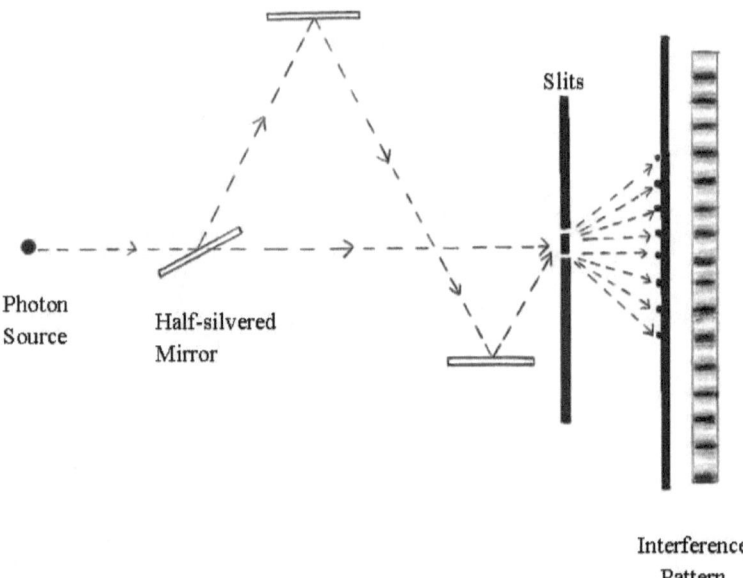

Modified dual-slit experiment demonstrating temporal nonlocality. Depending on which path it takes, a photon will pass through the slits and arrive at the photographic plate at different times.

The dual-slit experiment can be modified in such a way that one of the slits is closed electronically while the photon is in flight toward the apparatus or after it has passed through a slit but before it strikes the detector. In this so-called "delayed choice" experiment, the interference pattern disappears and only a simple diffraction pattern is observed. Theoretically, if the photon is forced to pass through only one slit, it is forbidden from acting like a wave and reverts to its particle nature. The same is true when the slit is closed after the photon passes through one of the slits, except that this effect precedes the cause (temporal nonlocality). However, the question is, how does the photon that goes through the open slit know that the other slit is closed or about to be closed after it passes through and that it must go to a different location on the photographic plate? Somehow, the photon is aware that the other slit is not fully available to it and it acts like a particle instead of a wave.

In another experiment, a distant quasar appears to be split into two objects by the bending of light from an intervening galaxy between Earth and the quasar. The light that is bent has roughly fifty thousand light years more distance to travel than the light that comes to Earth directly. However, the photon beams from the quasar interfere with each other in exactly the same way as if they were emitted seconds apart in the laboratory. It appears that the photons remain coupled despite the fact that they were emitted billions of years ago and arrive fifty thousand years apart.

Such experiments clearly demonstrate that the behavior of a quantum particle is not simply determined by the conditions of the test. The wavelike or particle-like behavior depends on the totality of the experimental apparatus and the intended observation of the experimenter.

This outlines an important aspect of quantum nonlocality, namely, that the observer and the observed system cannot be separated. The observer or his instruments are part of the system and influence the outcome of the observation. In other words, the act of observing alters or influences the system and this alteration occurs outside linear time. The observation can occur after the particle has passed through the slits, yet the way the system is observed still influences the outcome or behavior of the system. This is contrary to the common-sense notion of cause and effect.

To summarize what we know so far about quantum nonlocality:

- *The wavelike or particle-like behavior of a quantum particle is determined by the decision of the experimenter.*
- *An effect can precede a cause or even occur before an apparatus is turned on.*
- *The two complementary aspects of quantum particles, i.e. wave and particle, cannot be observed simultaneously.*
- *An "act of measurement or observation" is required before a quantum system can manifest in physical reality or become "real."*
- *Entanglement of quanta is nonlocal and occurs outside linear time.*

Quantum wave function

A very important part of quantum theory is the mathematical formulation that describes the possible states for quanta. Without getting into the complex math needed to get a true understanding of the quantum wave function, we will try to explain its salient features and why it has such importance in quantum physics and what it tells us about the nature of reality.

The pioneering work of Erwin Schrödinger during the early part of the twentieth century gave physicists a method for determining the possible wave functions for a system and how they evolve over time. Therefore, the wave function is also known as the Schrödinger wave equation or function. A wave function describes the properties of a wave such as water waves or vibrating violin strings. However, for quantum systems the wave function is not a wave in physical space, but a wave in an abstract "mathematical space." The Schrödinger wave equation provides an explanation for wave–particle duality. Before a quantum particle is observed and manifests in the "real" world it behaves like a wave. When it is observed or measured, it behaves like a particle.

In simplest terms, the wave function is a mathematical expression detailing all the possible states that a particle or system may take when it is "observed." This act of observation usually does not mean witnessing the quanta with the eyes or other sense organs but by instruments at the physicist's disposal. Since the wave function details possibilities, it speaks of a level of reality that is below what is considered "real." Physicists sometimes refer to this level as the "realm" or "domain" of the wave function. Any description of reality must include an explanation of this underlying and unseen realm from which "hard reality" emerges.

The wave function describes the probability that a quantum or system will assume a certain physical state when it is observed. Such observation causes the wave function, which can incorporate an almost infinite number of possibilities, to collapse into a single state. As mention earlier, this is termed the collapse of the wave function. The probability that observation will give any particular state is provided by simply squaring the wave equation for the system. Hence, the wave function provides the range of possibilities or potential states, and observation knocks it into just one state, but the likelihood that this state will manifest in physical reality is also predicted by the function.

Since we cannot know the state of a particle before it is measured, quantum theory concludes it must be a superposition of all possible states. Hence, the underlying realm of the wave function is not one of separate parts but one in which all possibilities coexist in a state of wholeness—everything being interconnected and interdependent. By necessity, this web of connectivity must permeate the entire universe and is a hallmark of nonlocality and wholeness, which is characteristic of the quantum realm.

Teleportation and quantum computers

Quantum teleportation is a process by which quantum information, such as the exact state of a photon, electron, ion, or atom, can be transmitted from one location to another. It is another well-established example of quantum entanglement. To understand this process as well as quantum computers, it is first necessary to introduce the concept of a quantum bit or "qubit." For the classical transfer of information in such things as computers or fiber optic cables, a simple two-state system is used, which is termed the "bit." A bit is represented by either a one or zero. The quantum analog of a bit is a qubit, which unlike classical bits confers more information than a one or zero. The quantum information encoded by a qubit contains information about the quantum state of the qubit—not just whether it is one or zero.

Quantum teleportation provides a mechanism for moving qubits from one location to another, without having to physically move the underlying qubit particle. Quantum teleportation can take place when there is previously established quantum entanglement between two quantum particles such as atoms at the sending and receiving locations, and the information about one of the atoms is sent by way of a "quantum channel" between the sending and receiving stations. Because such a channel must be set up using classical communication methods, the overall transfer of information cannot exceed the speed of light. In the process of transfer, the information carried by the atom at the sending station is destroyed. While the name teleportation conjures up images from Star Trek, it cannot be used to transport material objects—only information.

Quantum entanglement also makes possible the exciting new technology of quantum computing. Quantum computers use qubits instead of bits. Because a qubit can be a superposition of many states, the power of such computers can theoretically be orders of magnitude greater than that of classical computers in use today. The development of such "supercomputers" is still in its infancy but experiments have successfully demonstrated that simple computations can be carried out using a very small number of quantum bits.

In the future large-scale quantum computers will be able to solve certain problems, such as weather forecasting, much more quickly than today's classical computers. Unlike classical computers that can perform only one operation at a time, albeit very, very rapidly, the quantum computer utilizing qubits can perform many calculations simultaneously. For example, think of a single rat placed into a complicated maze with hundreds of dead ends and only one way out. It might take the rat many hours to find its way out as it tries numerous paths, only to be blocked most of the time. Now consider putting a hundred rats into the maze at the same time. Surely, one of the rats, by chance will find the elusive escape route in a short time. The quantum computer with its entangled qubits, which can be both zero and one and states in between, is able to make many calculations at the same time expanding its potential computational power millions of times over that of today's most powerful supercomputers.

Today the nonlocality of quantum particles (entanglement) is an accepted fact of nature that has been observed for not only photons but also electrons, atoms, and molecules. Physicists are dumbfounded to explain how it works but one day it will surely revolutionize the fields of electronics and communication. The connection that exists between distant but entangled particles through both time and space is perhaps the greatest mystery of quantum mechanics. Nonlocality in the physical realm is a fact. The best and perhaps only logical explanation for the phenomenon of quantum nonlocality is that all quanta are connected, inseparable parts of a wholeness that underlies our perceived reality.

4
Nonlocality of Time and Space

THE MICHELSON-MORLEY EXPERIMENT DISPROVED the idea of a "luminiferous aether" and showed that the speed of light was a constant and was not affected by the motion of Earth. This contradicted the commonsense notion that speeds should add up—i.e. a bullet fired forward from a fast-moving car should have a higher velocity than one fired backward. If light behaved similarly then one would expect that when Earth was moving toward a distant star, the light from the star would reach Earth more quickly than when Earth was moving away from the star. However, this is not the case—the light arrives at the same instant in both cases.

Albert Einstein realized that if the speed of light was a constant no matter what point of reference was used, then something else had to change to account for its constancy. He sensed that this "something" must be space itself. He proposed that space could flex and change, become compressed or expanded according to the relative motion of an object and an observer. The only constant was the speed of light itself or an integrated four-dimensional fabric he called space-time. These insights led to Einstein's special theory of relativity, which states that the universe has four dimensions. These are three of space—width, length, and height—and one of time. Time is not a separate dimension in this scheme, but is fully integrated with the three spatial dimensions. Hence, each of the four dimensions of space-time has a spatial and temporal component, which is required from the fact that both space and time are relative to the state of motion. Einstein theorized that with motion, space shrinks and time dilates, while for an object at rest, the movement through space-time is in time alone.

Einstein's equations indicated that the faster an object moves, the slower the passage of time and the more mass it gains. Ultimately, at the speed of light, time stops. However, for matter it would be impossible to attain this speed since it would require all the mass-energy of the universe. Subsequent experiments have proven Einstein's theories about space, time, energy, and mass to be correct. For example, the decay of an unstable subatomic particle that is accelerated near the speed of light in a particle accelerator is much longer than when it is stationary. Secondly, such particles gain the exact amount of mass predicted by the theory as they race in the accelerator near the speed of light.

However, photons, which carry electromagnetic radiation such as visible light, can move at the speed of light since they have no mass. Their internal clocks are stopped and they do not decay like other particles. As an object approaches the speed of light, both the object and space become compressed. For light, this would correspond to compressing space to a point. Light cannot move through space-time any faster than the speed of light since it is limited by the compressibility of space. Space can be compressed down to a point, but no further. Hence, not only is it impossible for any physical object to reach the speed of light, it is also impossible for light to exceed that speed.

Einstein's special theory of relativity also postulated the equivalence of mass and energy and gave us the famous equation that energy is equal to mass times the speed of light in a vacuum squared ($E=mc^2$). This equation demonstrates that matter is a condensed form of energy, and to this day physicists express the mass of subatomic particles in terms of energy—normally electron volts.

Einstein's second theory of relativity, termed the general theory of relativity, describes gravity as a geometric property of space-time. This theory predicts that gravity distorts space-time. The more massive the object the more it distorts or curves space. If the mass of an object is small then this curvature is minuscule and Einstein's equations describing how space-time is curved by mass become the same as Newton's equation describing gravity. The curvature of space-time caused by a massive object also causes light passing near that distant object to bend. This prediction has been verified experimentally, as has the existence of black holes—objects with such tremendous gravitational force that nothing can escape their pull, including light.

Several startling and unusual consequences arise from Einstein's new model of the universe. The most important, from our point of view, is the realization that four-dimensional space-time cannot change in time. All events that have occurred in the past or will happen in the future are already there. Space-time does not change in time nor can it change in space. It is simply there! Events are like seeds in a watermelon, positioned in fixed locations in the four dimensions of space-time. This means that our everyday experience of the flow of time from the past to the present to the future is actually an illusion.

This new picture of space and time is called "block time." Space-time is absolute and unchanging, much like Newton's three-dimensional space. There is the perception that things change in time, but in reality they cannot change within integrated space-time. Therefore, events do not take place in time; they simply are. The past, present, and future are all equally real and the flow of time is something human beings create as a convenient way to help them cope with their three-dimensional experience of reality. If one were to possess four-dimensional sight, one would experience reality far differently. Instead of seeing events unfolding with the passage of time, one would witness all time and space displayed in its totality. Theoretically, one could then witness events that took place millions of years ago, such as dinosaurs walking the earth, and simultaneously observe the future planet Earth being engulfed by our dying sun.

There are a number of weird implications that follow from this modern view of space and time:

- *Within integrated space-time, only events have meaning.* For example, if we witness the light from a supernova in a galaxy thirty million light years distant (the time light travels in one year), the event of the actual explosion thirty million years ago is consider simultaneous to the event of witnessing it today. The great distance separating an observer from the distant galaxy is simply one of the coordinates of space-time, which creates the illusion that much time has passed between the events, when in reality it is simply a measure of the space-time separating the galaxy and the observer.
- *Movement causes space to convert to time.* When an object is not moving, then it is moving in time alone. If an object is moving near the speed of light, then it is moving mostly through space and its clock will slow down relative to a stationary clock. For example, in the future we might develop a spaceship that can travel at 90 percent of the speed of light (270,000 kilometers/sec). After the astronauts reach full speed on their way to a planet circling a star twenty light years away, they will calculate the distance to the star to be only ten light years (because of the compression of space). They will then calculate that it should take eleven years to reach their destination. However, because their clock runs at half the speed as clocks on Earth, the event of their arrival after eleven years of their time will correspond precisely with the expected time of their arrival on Earth

(twenty-two years later). They would only age eleven years instead of twenty-two years, and upon their return to Earth, they may be the same age as a son or daughter they left behind on their journey.
- *A massive object distorts space-time.* Much as a bowling ball would bend a rubber membrane. This bending of space-time corresponds to gravity, and the distortion pulls on time as well. As a result, time is slowed down in a similar way to that of objects moving near the speed of light. This effect can be easily demonstrated by synchronizing two atomic clocks and moving one to a mountaintop for a week. Upon return to sea level the clock at high altitude is found to have run a little faster than the one left below because the force of gravity diminishes the farther one is from the surface of the earth. If astronauts were to orbit a black hole with its massive gravitational pull and then were able to break free and return to Earth, their clock would slow significantly relative to clocks on Earth. It is conceivable that after a few hours of "slowed" time near the black hole they could return to Earth and be younger than their grandchildren.
- *Spatial dimensions are compressed at high speed.* The shape of an object such as a starship would look compressed or flattened to someone observing it as it passed by Earth. To the astronauts on the starship, everything would look perfectly normal since everything including their measuring devices would have shrunk the same relative amount. This is termed the Lorentz–FitzGerald contraction, named after Hendrik Lorentz and George FitzGerald, who developed an equation to describe the effect.
- *Space and time are observer dependent.* Time and length may expand or shrink depending on the relative state of motion of the observer and the observed. As space shrinks, time expands (slows). Space is transformed into time and time into space. This is the hallmark of a four-dimensional substance in which each dimension has both a spatial and temporal aspect, which are fully integrated and inseparable.
- *Time is fluid.* The "now" is not the same for observers moving relative to one another. Time travel into the future is possible and even travel into the past is not ruled out.
- *Four-dimensional space-time is an unseen but underlying realm of reality.* From the three-dimensional perspective of human experience everything changes in time, but underneath this relative reality must exist an unchanging four-dimensional reality characterized by wholeness.

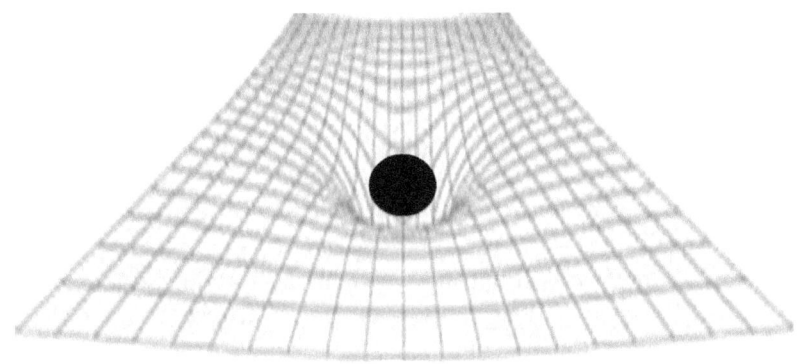

Illustration of how a massive object distorts space-time.

Human beings are incapable of witnessing all four dimensions of space-time. We are accustomed to living in three dimensions and perceive time flowing by. This is the working principle that we live by. Science says that being at rest relative to our surroundings, we are aware of moving in time alone, and for this reason the Newtonian picture of the universe works fine for us with space fixed in three dimensions and time flowing linearly. Nonetheless, what we take as the flow of time from the past to the present into the future is still an illusion—space-time is unchanging.

A new concept of time and now

The scientific view of space-time as a four-dimensional, unchanging "block" is very difficult for one to imagine with a three-dimensionally conditioned mind. An analogy that might help is how two-dimensional "Flatlanders" might experience the illusion of the flow of time if their world moved through a fixed three-dimensional world. The Flatlanders live on a plane and therefore only experience the edge of objects that reside on their plane; they cannot conceive of three-dimensional space. Now suppose their home plane moves at a constant speed through a three-dimensional space having a string of spheres lined up like a long necklace of pearls. They would witness the regular growth of a circle from a point, after which the circle begins to recede again to a point. They might associate this regular waxing and waning of the circle with time and set their clocks by this phenomenon in a manner similar to how early humans developed the calendar and time pieces to match the regular rhythms of the sun and moon. However, the flow of time these Flatlanders experience would be illusory. The flow of time they experience is simply the movement of their two-dimensional world through an unchanging three-dimensional space. Similarly, our experience of the flow of time is created by the fact that we are all moving at about the same speed relative to one another, and therefore our resultant movement through four-dimensional space-time is almost completely in time.

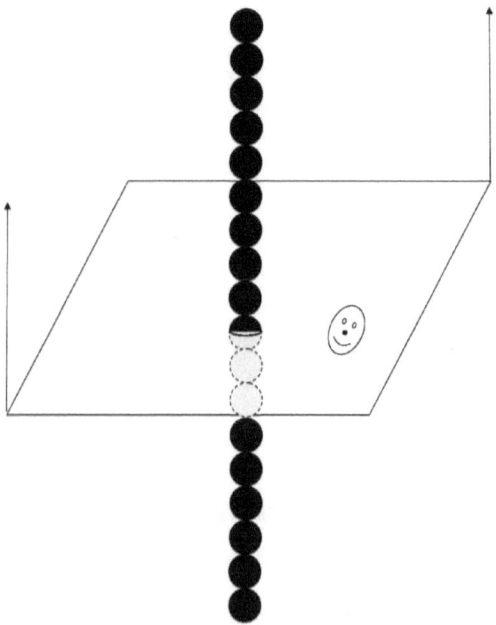

A plane moving at a constant rate through a string of spheres creates circles that grow and shrink regularly giving the illusion of time flowing to a Flatlander.

Another way to try to picture four-dimensional space-time is using a "timeline." Visualize a tube in three-dimensional space beginning with the Big Bang and extending out to eternity. A slice of this tube corresponds to "now" and a point on the plane sliced from the tube corresponds to a position in space.

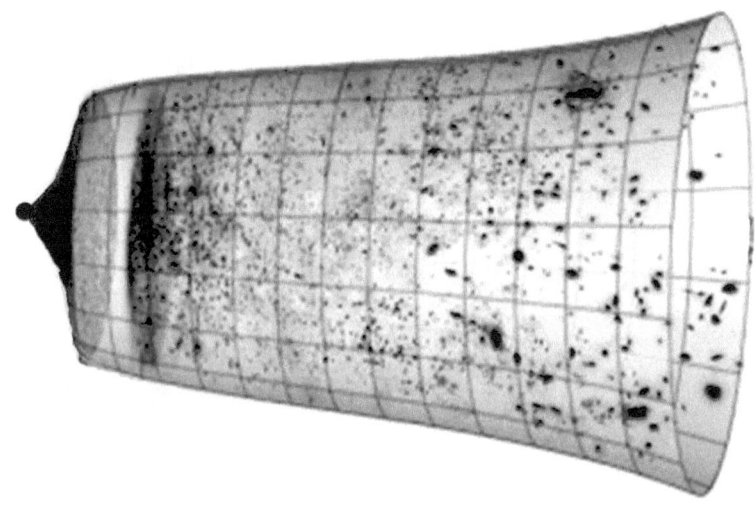

Illustration of a timeline beginning with the Big Bang

Actually, a slice of a four-dimensional space gives a three-dimensional space, not a plane, but this model of space-time is something that can be visualized. Next, consider what happens to this slice of the timeline, which may be called "now," if one is moving. Einstein's equations relating space and time say that if an object is moving, time slows down. Normally for observers on Earth who are not moving rapidly relative to one another, there is general agreement about now since the effect of motion on time is minuscule in this case. However, when two observers are moving relative to each other there is no longer complete agreement about now. It is as though the motion of one observer tilts the knife and he obtains a different slice of now. For example, this effect is observed when two atomic clocks are synchronized and one is flown around the globe in a jet. The clock in motion is found to have slowed down slightly—just as predicted by theory. Its now is slightly different from that of the stationary clock.

This effect is minute at the speed that a jet travels, but if astronauts on a spaceship a great distance away were moving in a direction away from Earth, their slice of the time-line would be angled toward the past. Their clock would be slowed down relative to those on Earth, and any information they receive from Earth using regular signals, limited by the speed of light, would be angled toward the past. Even if the angle were very small, the effect would be magnified because of their great distance from Earth. "Now" to them would include events that occurred in Earth's past, such as the collision of an asteroid with the moon that already took place. On the other hand, if the distant astronauts were moving toward Earth, their slice of the time-line, and hence their now, would be angled toward the future. They could witness events that have not yet occurred on Earth. All these slices of the time-line are equally valid and real. Each person's perception of now is equally valid no matter whether that person is in motion or not. According to physics, all time is out there. All time exists simultaneously in space-time, and the now, just like the past and future, are observer dependent and therefore mutable.

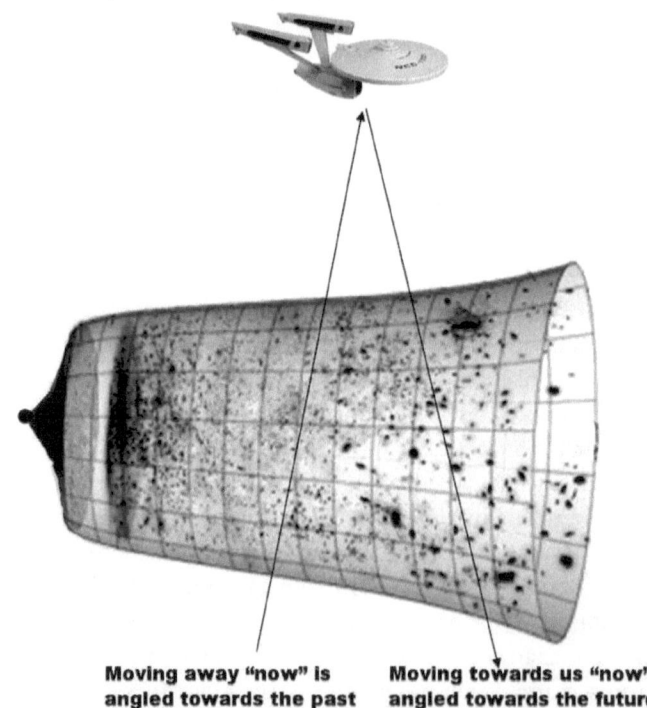

Moving away "now" is angled towards the past **Moving towards us "now" is angled towards the future**

An observer in motion relative to Earth gets a different

slice of the time-line and consequently experiences a different "now."

An important question is what happens to the idea of free will if future events are completely determined. Can there be free will if the future is already "written"? The answer is that humans make decisions on how to act and do not know how events will unfold. It is though they were actors in a movie and do not know the script. For them every moment is new and they decide how to react to life's events. For example, a windstorm damages your house and that of your neighbors. You are free to retreat in self-pity or grow by helping others recover from the disaster. Humans have total freedom as to how they focus their mind and direct their energies. On the other hand, the director of the movie, God, knows exactly how the script will be played out. Everything takes place within his field of knowledge (mind) in the eternity of the now.

The arrow of time

Physics does not define or require time in most of its equations since they work equally well without a factor for time, and whether it flows forward or backward. Relativity theory describes time in terms of block time in which time is mutable but space-time is not. If space-time is really frozen then the perception of past, present, and future is an illusion, but the question is why time is universally perceived to flow in only one direction. In other words, does the arrow of time have any basis in science, or is this flow of time merely a construct of a three-dimensional mind? The experience of time flowing is similar to the constant changes that occur in the frames of a moving picture. Certainly, there must be a basis in the laws of physics for the fact that this "moving picture" that is experienced only moves in one direction. The answer to why time's arrow only points in one direction seems to lie in the concept of entropy.

Entropy is a thermodynamic property that is needed to explain certain reactions that occur spontaneously without the expenditure of normal energy such as heat. For example, consider two containers: one filled with oxygen and the other with nitrogen. Now connect the containers by a small tube. Gradually oxygen molecules from one container will pass into the other container while nitrogen molecules will flow in the opposite direction. Eventually, the two gases will become completely mixed. The driving force for this mixing is called entropy. The system goes from a highly ordered state (two pure gases) to a less ordered state (mixed gases). Hence, entropy is the cause of greater disorder or randomness, and it is ever increasing in the universe.

The everyday experience of the flow of time is tied inexorably to the constant increase in entropy that is observed. For example, consider a two-hundred-page manuscript all neatly ordered from the first to the last page. Now throw the papers into the air. It is unlikely that all the pages will return to their original order. This is because there are millions of possible ways the papers could fall in a disordered fashion and only one way they could return to their original state. The driving force behind the tendency for a system to go from order to disorder is entropy.

One might assume that living organisms violate the rule that entropy is always on the increase. For example, human beings are able to construct many well-ordered things starting from disordered raw materials. However, living organisms do not actually violate the second law of thermodynamics, which states that increasing entropy helps drive change in most everything. When the results of the metabolism of highly ordered food molecules are taken into account, the net effect is still an increase in entropy. Ultimately, all the energies utilized by living organisms can be traced back to the energy of the sun, which, by converting mostly hydrogen into heavier elements and lots of energy, contributes to an increase of entropy in the universe.

Entropy is always increasing. Disorder has increased from the time of the Big Bang. If the time-line for the universe is put in reverse, there would be an increase in order toward the past, culminating in a point of maximum order at the initial point of the Big Bang. How or why this came about is a mystery to cosmologists. Since there was no space-time before the Big Bang, the time-line starts there in a state of maximum order. Accordingly, what we call time can only move in one direction coinciding with the flow from order to disorder.

Time travel

Science fiction is rich with stories of time travel—both to the future and to the past. But is time travel really possible? Physics does not rule it out and the new understanding of space-time provides scenarios where it may be possible in the future. It is known that motion affects time, and that the faster one travels, the slower a clock ticks relative to a stationary clock. It is also known that gravity distorts space-time and has a similar effect on the passage of time. Thus travel into the future is quite possible. All that is needed is an advanced propulsion system that could propel a vehicle to close to the speed of light. This is no easy task, since at present our fastest space vehicles go only about 0.004 percent of the speed of light and relativistic effects do not become significant until one reaches at least 50 percent of the speed of light. Moreover, getting close enough to a black hole to slow the clock in a spaceship would mean that the enormous gravitational force of the black hole would increase an astronaut's weight more than a thousand fold. This could literally tear their body apart, not to mention the tremendous difficulty they would have in escaping the intense gravitational pull of the black hole.

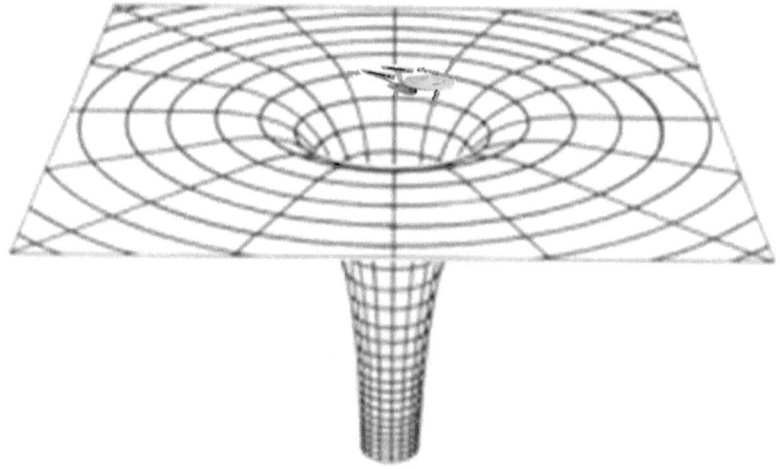

A spaceship circling the event horizon of a black hole would be subjected to tremendous gravitational force and would find it very difficult to escape from the black hole's gravitational attraction.

Travel to the past might prove to be more difficult. Theoretically, the science of relativity teaches that as an object approaches the speed of light its clock slows down, and at the speed of light the clock stops entirely. Travel faster than the speed of light and the clock would begin to move backward. Hence, if there was a way for humans in the distant future to develop the technology (like the warp drive of Star Trek fame) to travel faster than the speed of light, then they could turn back their clock, so to speak, and return to Earth in the past. However, relativity theory precludes this from occurring since no object possessing mass can ever attain, much less exceed the speed of light.

Another possibility for traveling in time is to find a wormhole that connects one part of space-time to another. Using a wormhole, one could conceivably travel into the future or to the past. If astronauts find a wormhole that takes them to the past, they might return to Earth in the past. They could conceivably meet themselves and discuss the future. The laws of physics do not rule out such a scenario. However, it is highly unlikely that any physical object could survive such a journey because physics says that matter with negative density is needed to pass through a wormhole and no such material is known to exist in the universe. Moreover, if time travel to the past were possible, it would create impossible paradoxes. For example, you could meet your grandfather in the past and murder him before he had children. In this case, you would never have been born and therefore could not possibly kill him. In addition, if time travel to the past were possible, then we would expect to be constantly bothered by visitors from the future who would no doubt want to come and observe human history firsthand.

Relativity theory vs. quantum theory

Quantum mechanics is extremely successful for describing the behavior of minute bits of matter and energy (quanta) but it has nothing to say about the large-scale features of the universe. It describes reality as discontinuous, and for most interactions on the macro scale, the quantum of action is so small that it can be ignored. On the other hand, relativity theory is wonderful for describing the large-scale features of the universe but fails to work at the minute level. It appears to describe a continuous reality, and for most situations, speeds are so much slower than light that relativistic effects can be ignored. It is almost as if these two great theories describing reality were incompatible.

Scientists are now proposing that there might be more overlap in these two theories than previously considered. For example, space-time can be considered discontinuous. In 1913, Max Planck established a relationship between the speed of light (c), quanta (h), and the gravitational constant (G). This gave him minimum values for time (5.3×10^{-44} seconds), length (1.6×10^{-33} cm), and mass (2.2×10^{-5} g). This would mean that within space-time, time fluctuates 2×10^{43} times per second and the size of a proton is about 10^{17} times the size of the tiniest bit of space. Unfortunately times and sizes on this scale are too small to measure, so the theory that space-time is discontinuous remains theoretical, but this theory would explain why space-time contains so much energy.

A second overlap in the two theories is the role of the observer. In the quantum mechanical description of matter-energy there must be observation before the wave function can collapse. The measuring apparatus is considered part of the quantum mechanical system, but ultimately an act of consciousness or observation is needed before the system can assume physical reality. Similarly, in relativity the observer is required, with particular space-time coordinates. A full description of either a quantum mechanical or relative system depends on an observer.

In relativity theory, time is fully integrated with space and it does not change or flow within integrated space-time. Events take place within space-time but the concepts of before and after or cause and effect have no meaning in the context of the wholeness of space-time. Similarly, quantum theory is also timeless. The wave function is a timeless domain that does not have a time direction. It is not until the wave function collapses that an arrow of time is introduced. In other words, observation is needed for the collapse and it creates the arrow of time. There is no better example of this than the delayed-choice experiment in which observation of a past event is influenced by choices made in the present.

Finally, complementarity plays an important role in both theories. Quantum physics has complementary constructs, such as wave-particle, position-momentum, field-gauge particle, part-whole. In relativity theory, Einstein showed that mass and energy as well as space and time were complementary. As mentioned before, when complementary aspects are needed for a full description of something, then the true reality of that thing must lie at a deeper level than that of the two complementary descriptions.

Summary

Einstein's special theory of relativity with its mathematical description of a four-dimensional space-time continuum has been verified by numerous experimental observations. Furthermore, predictions made by the theory have proven correct and highly accurate. It is one of the most important scientific discoveries made in the last thousand years. Many of today's technological advances (e.g., the GPS system) depend on the relative mechanics that are derived from the theory. However, what it says about the nature of reality is both mindboggling and revolutionary. It says that beneath this ever-changing realm of human experience lies a deeper, singular and unchanging realm of reality. The implications of this model of the universe are enormous, and most scientists today do not seem to comprehend what this model of the universe says about reality. One might even wonder if Einstein understood the implications of his discovery for scientific epistemology. It is as though there is an "elephant in the room," and scientists choose not to see it. Perhaps this is because it describes an all-encompassing timeless domain of which our three-dimensional world is merely a shadow. This speaks of an "other worldly" or metaphysical view of reality similar to that proposed by Plato and many other philosophers and prophets; if true, this would thoroughly destroy the materialistic worldview that most scientists embrace.

As Earthlings, we do not move very fast relative to one another. Consequently, we all agree on the same now. Our movement through four-dimensional space-time is in time alone. Events appear to unfold before us like the frames of a moving picture. However, what is actually creating the illusion that time flows is that our three-dimensional world is moving through a four-dimensional reality at a constant speed, and things appear to change in time. On the other hand, God possesses four-dimensional "sight." He can witness the creation in its entirety, which means that he is witness to all space and time in what could be called "the eternal now."

Quantum mechanics provides the same view of reality. Quanta born of the same source are coupled with one another, and this connection is not limited by either space or time. In fact, all quanta were entangled in the beginning of the universe, and such entanglement grows exponentially in proportion to the number of particles involved in the original quantum state. Quanta are connected precisely because they emerge from a timeless, hidden domain in which everything is connected. Hence, the entire universe is composed of a network of particles, all of which are in contact with one another, and they leave the hidden domain of wholeness to enter physical reality only when mind or consciousness gets involved. Without one of these, whether it is individual mind/consciousness or Cosmic Mind/Consciousness, nothing could come into concrete existence. In the final analysis, science has uncovered an incredible, almost incomprehensible cosmic dance in which nonlocality is the rule, not the exception, and the universe is described by wholeness instead of individual parts. These basic principles are entirely consistent with spiritual ideology. Only a monistic or spiritual worldview can offer a satisfying explanation for all these confounding facts about reality. We will explore in greater depth the logic behind this statement in Part III of this book.

PART II

NONLOCAL MIND

The idea that mind is simply a product of the brain is the hypothesis embraced by a majority of today's neuroscientists. Much of the research into brain anatomy and function in the last few decades has reinforced this doctrine. New methodologies based on behavioral, clinical, pharmacologic, genetic, neurosurgical, and electrophysiological probes as well as neuroimaging techniques, have increasingly demonstrated the close linkage between brain physiology and mental states. This has convinced most neuroscientists that almost all mental states have a physiological basis and that there is no longer any need to consider the dualistic separation of mind and brain.

It is true that there exists a strong correlation in animals between the evolution of the brain and nervous system and higher mental faculties, and that damage or changes in the brain cause changes in mental functioning and consciousness. However, it is not true that correlation is causation.[1] No doubt it has been easier for neuroscientists to study how changes in the brain affect mental states. However, the mind-brain equivalence begins to break down when the study switches to how a change in a mental state (mind) affects a physical state (brain).

It is our feeling that just like eighteenth- and nineteenth-century scientists who believed that the existing laws of classical physics could explain everything, certain anomalies in the theory that mind equals brain exist that disprove this materialistic doctrine. As a result, one must seriously consider the alternative hypothesis that mind is much more than the brain and that the correct ontology is top down, in which mind and matter are epiphenomena of Consciousness.

In this section, we will explore much of the empirical data that challenges the theory that mind is solely a product of the brain.

5
Why Mind is not Brain

THERE ARE NUMEROUS PROBLEMS with the doctrine that mind results from the neurophysiological, i.e. electrochemical events and processes occurring in the brain, not the least of which is the failure to locate any physical structure in the brain that is responsible for conscious self-awareness. If there were a brain basis for the sense of self then it makes sense that damage to some part of the brain would destroy this; however, this is not observed. Consciousness and a sense of self seems to survive such insults (although diminished in many cases), even when large sections of the brain are removed or damaged.

Many other mental faculties and phenomena lend strong support to the theory that mind is not simply the output of the physical brain. Some of these include: a unified sense of self, placebo effects, stigmata, hypnotically produced physical symptoms, memory, genius and creativity, mystical experience, remembrances of past lives, out-of-body consciousness, near-death experiences, and finally psychic phenomena (ESP).

We shall take up each of these "anomalies" separately. Each seems to contradict the doctrine that mind is a product of matter. Hopefully, when all is said and done the reader will agree that there is a preponderance of evidence indicating that mind is not simply a product of brain and that it is a subtle, nonlocal faculty with far greater capabilities than the physical brain.

The unity of self-awareness

At the microscopic level, a person is not the same as they were even a few weeks ago. There is a continuous flow of atoms into and out of the body. Much like water flowing past a bridge, which is changing continuously, the same can be said of the material making up the body and brain. After only a couple of weeks, the carbon, hydrogen, oxygen, and nitrogen atoms that make up the body are replaced with new atoms. Iron is probably the longest-lived atom in the body and half of it is replaced after six months. Hence, the material composing the physical body is constantly changing. The same could be said about the physical structures or anatomy of the brain. The brain one has today is nothing like the one they had as a newborn or as a child. The brain, like other organs of the body, undergoes constant change. Yet there is something about a person that is the same since they were born. That something is the sense of self-identity. Physically one may have little resemblance to when they were a child, but they are aware that they are the same person at age sixty as they were at age six. The continuity of self-awareness does not seem to depend on the continuity of the component parts or any physical structure.

The unity of experience is very difficult to explain using the theory that brain is the basis for mind and consciousness. It is now known that sensory inputs are handled by different mechanisms and/or anatomically separate parts of the brain. For example, visual stimuli, which include color, form, and motion, are handled by different parts of the brain; yet somehow one has a unified visual experience. The unity of sensory experience has no identifiable anatomical basis and therefore must be achieved by some other means than by brain anatomy. If mind were solely a product of brain, then for there to be unity of experience there would have to be what is termed "anatomical convergence." Such convergence does not exist so the prevailing materialistic theory involves large-scale gamma-band oscillatory electrical activity in the brain.[1] The problem with this theory is that there is considerable evidence that the unity of conscious experience continues during cardiac arrest and other stoppages of blood flow to the brain, in the form of out-of-body experiences (OBEs) and near-death experiences (NDEs). Even when there is no measurable electrophysiological activity of the brain, the faculty of self-awareness and unity of experience continues. Thus the experience of being a conscious, singular person cannot be adequately explained by the physicalist doctrine that mind is merely a function of brain activity.

Mind as a computer

With the development of modern-day computers, it has become popular to equate the brain with a supercomputer. Advances in artificial intelligence

(AI) have made it next to impossible to determine whether one is communicating with a computer or a real person, but this is very different from saying that computers will one day attain understanding and consciousness. For example, a computer can easily translate this paragraph into Japanese by manipulating symbols and words, but could it be programed to learn the meaning of what it has translated? Critics of this computational model of mind say no; it is impossible for a machine to gain understanding and even more difficult to conceive of how it could obtain conscious awareness.[2]

Psychosomatic illness and how mind affects health

It has been known for many years that mental states affect the body, and doctors are now taught that symptoms of disease are sometimes only in the "mind" of the patient. Psychological medical approaches, such as alternative medical treatment, a hands-on approach by a doctor or healer, and the use of a placebo may be particularly effective in bringing about a cure of such patients. Psychological feelings of hopelessness and depression are strongly correlated with an increased risk of chronic diseases such as heart disease and cancer. On the flip side, experiences of joy and laughter have been demonstrated to improve health. Studies of how mental states affect the immune system (psychoneuroimmunology) offer one possible mechanism by which one's mental state might affect their health. There is now convincing evidence that the brain and nervous system are connected to the immune system and to the release of "stress hormones" such as cortisol. Because of the myriad connections between the brain and the body, neuroscientists today talk in terms of "mind-body unity," but what they really mean is "brain-body" unity.

There are numerous other examples of how mental states affect a physiological response or the health of an individual. For example, studies indicate that there is increased mortality following bereavement. The stress and sorrow following the death of a loved one may cause a person to give up all hope and quickly die from cardiac arrest. In some cultures, the fear from receiving a curse has similarly caused sudden death. On the other hand, there have been numerous reports of individuals that postponed their death until after some significant event such as the birth of a grandchild. Research has shown that there is a positive correlation between positive emotions and health, and that such things as meditation, imagery, biofeedback, relaxation training, and hypnosis are effective for improving disease states.[3]

The placebo (and its counterpart the nocebo) effect is fully recognized as a way in which mental expectation can affect pain and illness. The administration of a placebo, which by definition has no physical effect on the body, creates a psychological effect, which in turn can cause a measureable physiological effect. The placebo effect is therefore opposite to the physicalist doctrine that mental states are a result of physiological changes. The placebo effect is so well established in modern medicine that it is now required to include a placebo in clinical studies designed to show the efficacy of medical treatments. Not only is the patient unaware whether they are receiving the experimental drug or treatment, in most studies the persons administering the treatment are also "blind" to whether the patient is receiving the experimental drug or a placebo. This is because if the doctor or other healthcare professional knows whether they are administering a placebo or not, they might influence the patient outcome in subtle ways. Because patients often show improvement in "double blind" studies of this type, many experimental drugs and treatments have failed their clinical trials because there was no significant difference between the placebo and experimental groups.

Other mentally induced physiological changes

One of the more curious effects of this type is stigmata. This is where a person develops marks and sometimes bleeding wounds that Christ was thought to have suffered during his crucifixion. There have been hundreds of such cases reported in both the common and medical literature. One of the first such cases was that of St. Francis of Assisi. Persons with this affliction are most often intensely religious and have become emotionally tied to the suffering of Christ. Similar effects have also been observed in a nonreligious setting in which a person has bled from their hands, armpits, and eyes while undergoing strong emotional stress.[4] There has been no satisfactory explanation of how the brain and nervous system could produce such localized skin responses.

Another example of mind affecting physiology is false pregnancy in which a woman believes she is pregnant, but is not, and displays many of the typical signs of pregnancy. Symptoms may include abdominal enlargement, cessation of menstruation, breast changes, morning sickness, the sensation of fetal movements, and labor pains. Reports of this type of mental disturbance were more common in the past before the development of modern diagnostic techniques.

There have also been numerous reports of persons who almost overnight suffered whitening of their hair or skin in response to a severe fright or emotional stress. There is no known biological mechanism that could produce such changes, especially in hair, which except for the root is composed of nonliving tissue and is therefore not subject to physiological changes in the body.

It is well known that hypnosis in some individuals can produce surgical analgesia, changes in allergic reactions, and changes in autonomic functions such as heart rate, skin temperature, blood glucose, salivation, etc. A less known effect is on skin markings and blisters. Apparently, changes in the skin are particularly susceptible to suggestion. Conditions such as warts, eczema, psoriasis, and fish-skin disease (ichthyosis) have been cured by hypnosis.[5] Hypnosis can also induce skin conditions such as bruising, redness, blistering, and bleeding at specific sites on the body suggested by the hypnotist. Again, there is no satisfactory explanation of how such specific physiologic effects can be caused by brain-body connections.

A few exceptional individuals have been studied that display what is termed "skin-writing." One such person was Olga Kahl, who produced on her skin in less than a minute a mentally communicated word or image. Her case was extensively studied and it was shown that the red color of the "writing" was well below the surface of the skin and required an exceptional control of the peripheral circulation.[6] Neuroscientists are unable to offer a biological explanation for how someone could have such exquisite control over capillary blood flow.

Studies have shown that trained meditators and yogis may have extraordinary control over otherwise autonomic processes. Examples include imperviousness to pain or cold, changes in skin temperature, and slowing or stoppage of heart and lungs. One such example taken from the medical literature is that of yogi Satyamurti. He was described as a thinly built man of about sixty years of age who volunteered to be confined to a small underground pit for a period of eight days. He said he would enter a state of samadhi (absorption in the Divine) and would not need any food, water, or air during the eight days, and stipulated that no matter what happened, the pit should not be opened prematurely. The investigators fitted him with a 12-lead EKG and soon after the pit was sealed his normal heart rate began racing, eventually reaching 250 beats per minute, whereupon it stopped beating. The straight line on the EKG persisted for 7 days after which normal sinus-heart rhythm returned. Upon leaving the pit, Satyamurti's body temperature was found to be only 94.60 F. The investigators saw no tell-tail signs of an electrical disturbance, either before his heart stopped or after it restarted, which would have occurred if he had removed the leads.[7]

Transpersonal influences

One could argue that the psychophysiological effects outlined above are simply a reflection of the unity of brain and body. This is far from a satisfying explanation since it does not say how the brain and nervous system cause such specific effects. Furthermore, this explanation breaks down completely for the many examples in which the mental state of one person influences the body of another person, sometimes without their knowledge of the event or condition. For example, a man develops a severe pain in the temple at about the same time that a relative shoots himself in the temple but before news of the event reaches him.

Sometimes the effect on another person is purely mental. Ian Stevenson published a review and analysis of 160 such cases in which a person had a strong impression about something happening to another person who was far away.[8] This would appear to be an example of extrasensory perception (ESP), but materialists would reject this hypothesis because according to materialistic doctrine, it is impossible. However, they offer no other plausible explanation except to say such reports are hoaxes.

Multiple Personality Disorder (MPD) might also fall under this category. It is characterized by the appearance of at least two distinct and relatively long-lasting identities or dissociated personality states that alternately control a person's behavior. Often there is impaired memory of what took place in one personality while in the other personality. What is more interesting is the fact that the different personalities may exhibit different physiologic conditions. For example, one personality may be allergic to a food and the other not. One may be right handed and the other left handed. One may have diabetes and require insulin and the other does not; or one requires glasses and the other does not.[9]

There is also considerable evidence that prayer, with or without a person's knowledge, improves medical outcomes. Larry Dossey conducted ten years of research on the relationship between prayer and healing and found compelling evidence that it can complement modern medical treatments. The scientific evidence for this is outlined in his book *Healing Words*.[10]

Memory

Some neuroscientist would have us believe that memory can be completely explained by electrochemical processes in the brain. However, the evidence clearly indicates that memory is a very complex process that involves both the brain and the nonphysical entity we call mind. Research has shown that memories are not stored in any specific location of the brain but are spread throughout the brain much like a hologram stores a three-dimensional picture throughout its entire matrix. Memories may consist of exquisite details and mental pictures that one is not even conscious of most of the time. These detailed recollections may be brought out by electrical stimulation of the brain or by hypnosis. For example, a person may recall under hypnosis a detailed description of a criminal perpetrator, including details of his tattoo, earrings, scar, and even the license plate of his get-away car, details they had no recollection of following the incident. A very small minority of people have exceptional memory (hyperthymesia). Given any date they can recall with great accuracy events that they experienced on that day.

One of the first theories for memory postulated by neuroscientists was the "trace" model. Experience causes physical changes in the brain, which can later be recalled when these physical traces or "engrams" are retraced cognitively. This model seems to work fine for simple creatures such as slugs that clearly undergo changes in their primitive nervous system in response to changes in their environment. However, such learned responses are nothing like human memory with its ability to recall at will details of past events (autobiographical memory) or general knowledge (semantic memory). In addition, neuroscientists have been unable to locate or identify any engrams in the brains of test animals. Engrams, if they exist, would be physical pathways that carry information about a past event or learned bit of knowledge. Instead, research shows that memories are not localized in any particular part of the brain. For example, in experiments using rats that were trained to run a maze, tissue was removed from their cerebral cortices before re-introducing them to the maze, to see how their memory was affected. As increasing amounts of tissue were removed memory was degraded, but remarkably, it made no difference where in the brain the tissue was removed.[11]

Another problem with this model is how memory is experienced. Most memories are not simply replays in the mind of the original event. Instead, persons will often see themselves witnessing or taking part in an event. People are aware that they saw or did something. In other words, there is self-awareness, which is a mental function, not simply a replaying of events. Along these same lines is the question of how memory is placed in time. Somehow, the mind "places" the memory in a personal time-line in a way that it relates to other dates, names, events, etc. Thus, instead of recalling a specific memory or impression, most memories are associated with a slew of other memories and general knowledge, all of which must lie in separate physical structures according to this theory. In addition, there is the question of what stamps a memory as genuine rather than imagined. It cannot be the vividness or intensity of the experience, since hallucinations may be equally or even more intense than actual experiences. People clearly possess a higher mental function that can be called personal awareness or consciousness that allows them to place memories within the greater context of experience and decide whether they are real or imagined.

A new and more viable theory of how the brain stores memories is the network model. A typical human brain weighs about three pounds but contains a trillion cells, 10 percent of which are nerve cells; hence, they number about one hundred billion. On the average, each neuron can receive signals and be connected to about five thousand other neurons. Hence, the total number of possible connections for this neuro network is ten to the millionth power.[12] This number is far greater than the number of atoms in the universe. According to this model, memories and learning are distributed across many neurons and their connections or pathways, resulting in a network that continues to function even when parts are damaged or removed. The brain functions like a super supercomputer, and like a computer with tons of memory chips, personal memories and knowledge are "stored" in the neuro network much like files on a computer. This hypothesis is fine, but can it work without the nonphysical entity we call mind? I believe the answer to this question is no, because in order for the data stored in the physical brain to be looked at or retrieved we would need a second memory system or "self" (sometimes termed a homunculus) that already possesses the skill and memories needed for the retrieval.[13] The existence of a witnessing entity for the brain or a "self" is antithetical to the brain-equals-mind theory because it implies the existence of a unitary, nonphysical mental function. Additionally, in Chapter 7, we will look at evidence for memory of past lives or noncerebral memory, which precludes the possibility that mind is brain.

Why quantum mechanics teaches that mind is not brain

So far, the arguments for why mind could not simply be a byproduct of brain are based on the occurrence of mental functions or states that do not have a simple psychophysiological explanation. However, quantum theory also makes a very strong argument that there must be a nonphysical entity present before any particle or object can become "real." A quantum mechanical wave function describes not only quanta (e.g. an electron), but can be used to describe any physical object including the brain. Quantum mechanics teaches that the quantum wave function describes the potential states that an object may assume when it is observed or measured. Prior to such observation, that object or system remains in potentia and has no physical reality. Since all physical reality must be considered quantum mechanical, what can collapse the wave function? It cannot be a brain-based observing apparatus because the brain is also a physical system that has its own wave function. Any brain-based observing system must be considered part of the brain's wave function. This means that something nonphysical is needed to collapse the wave function of the brain and bring things into physical reality. In other words, the brain is just like any other quantum system requiring an observer that is not part of the whole system before it can leave the realm of the wave function (potentia) and assume physical actuality. The only thing that could bring about this "collapse" is mind. Mind is nonphysical and therefore not part of the quantum mechanical brain. Hence, quantum theory also argues strongly that mind and consciousness are primary and matter/brain are secondary phenomena. Such a top-down view of reality is also consistent with other mental phenomena that are clearly nonlocal in nature. We shall explore several of these in the next few chapters.

The quantum mechanical model for the relationship between mind and brain also solves the problem of how something nonphysical (mind) can affect something physical (body). Consider for example the act of raising an arm. There is mental intention that causes nerves to fire, which causes muscles to contract, and the arm is raised. One of the chief arguments made by skeptics of the top-down worldview is that mind, which is non-physical by definition, could in no way affect anything physical without the exchange of energy—and no such interaction is therefore possible or scientifically conceivable. However, the skeptics would be wrong. Such a mechanism does exist (i.e. observation), and it is a necessary and sufficient process for bringing any quantum mechanical system into actuality. Hence, awareness or consciousness is the nonphysical ingredient needed before the brain can actualize the firing of a single neuron. Intention ("I do") is translated into physical action when it causes the wave function of specific brain neurons that are needed to move the arm to "collapse" into a specific physical state that fires the required neurons. No exchange of energy is needed—only intention or conscious awareness of what one wants to do. In this way, consciousness becomes the all-important ingredient needed to bring potentiality into actuality. Hence modern science, and quantum mechanics in particular, offers an important "proof" that downward causation is the only rational explanation for reality. In the next few chapters, we will investigate several other phenomena that reinforce this idea.

6
Mystical Experiences

> I appear before a person according to his or her desires. His or her whole being will be filled with My being. All the jivas (units) of this universe are rushing toward Me, knowingly or unknowingly. This is the final secret of the universe.
>
> —Lord Krishna, Bhagavad Gita

PSYCHOLOGISTS RECOGNIZE THREE STATES of consciousness: waking or normal consciousness, subconscious or dream state, and deep, dreamless sleep. Mystics on the other hand claim that there is a fourth state of consciousness in which one experiences the unity with all things. This transcendental state of awareness is said to be as different from normal waking consciousness as the dream state is from normal consciousness. The fourth or mystical state of awareness is sometimes called cosmic consciousness.

One of the salient features of a mystical experience is its ineffability or inability to be described. Another feature is the experience of knowledge that is nonlocal, intuitive, and not at all intellectual in nature. Experiencers feel certain that they have been witness to a higher reality than that of everyday life. Additionally, such experiences are typically short-lived and may be spontaneous or result from intense devotional sentiment or meditation. People report that the experience, no matter how transient, led to a profound change in how they view and conduct their lives—a change that lasts for the rest of their lifetime.

What sets the mystical experience apart from other psychological states, such as hypnosis, hysteria, hallucinations, etc., is the remarkable similarity of the experience despite historical, geographical, cultural, and religious differences. One of the common characteristics of the experience is the feeling of unity with the cosmos or God. One's consciousness is not tied to the body but seems to merge or become one with the whole of creation. Secondly, the flow of time seems to cease and one is aware of the timelessness of Ultimate Reality—everything exists and moves in the eternal now. The perception that time flows from the past to the present to the future is seen as an illusory aspect of the relative reality of everyday consciousness. Even individuals with a strongly religious background describe the experience as cosmic rather than religious. They might describe feeling that they were in the presence of their chosen savior, prophet, guru, etc., but in the full-blown mystical experience qualifications based on religious preferences disappear into the oneness of being. These common features of the mystical experience are seen despite differences in the age, culture, nationality, religion, and gender of the experiencers. This and the fact that the experience has such a profound and life-changing effect provides strong evidence that the experience is not illusory but a vision of a higher reality.

Most of the world's great religions grew from the mystical visions of such men as Sadashiva, Lao Tzu, Moses, Buddha, Jesus, Mohammed, and Baha'u'llah. Although the original mystical message of such prophets may have been diluted or obscured by centuries of religious doctrine and rituals, certain core mystical traditions have survived. These include the yogic and Tantric practices of Hinduism, Buddhist meditative practices, Kabbalistic Judaism, Christian mysticism, and Islamic Sufism. The goal of all of these religious traditions is the same—for the individual to undergo a mystical union with Cosmic Consciousness or God. We find different terminologies used to describe this merger of the individual with the Cosmic but the meaning is the same whether it is called enlightenment, self-realization, liberation, salvation, samadhi, nirvana, moksha, mukti, or satori.

One of the richest mystical traditions is yoga. Yoga means "union," and the person most responsible for systematizing yoga was Patanjali, who lived about 150 BCE. Patanjali compiled the Yoga Sutras, which delineated Ashtanga Yoga or the eight-limbed path of yoga. These eight limbs were *yama* and *niyama* (do's and don'ts, or ethics), asanas (postures of hatha yoga), *pranayama* (breath control), *pratyahara* (withdrawal of mind), *dharana* (concentration), *dhyana* (meditation), and finally samadhi (suspension of mind and absorption in the One). The purpose of the yogic practices is to decondition the mind, to free it from cultural, psychological, and ego-related bondages. The mind is thus free to associate with the Higher Self or Cosmic Entity. The mystical union of the little self (ego) with the Higher Self or Cosmic Consciousness is the ultimate goal of yoga. The analogy is that of a river running into the sea. It no longer remains a river—it becomes one with the sea.

Buddhist meditative practices attempt to do the same thing. The practitioner's mind is progressively cleared of preconditioned beliefs and assumptions about life as it moves toward a deconditioned state of pure consciousness, emptiness, or extinguishment of mind (nirvana).

A few examples of people's feeble attempts to describe their mystical experience may help shed some light on the subject.

St. Catherine of Siena (1347-1380)

At the age of sixteen, Catherine joined the St. Dominic nunnery in Siena, Italy and developed a gift for contemplation of the Lord. She frequently experienced ecstatic union with God. One such rapture is described in her biography.[1]

> On the feast day of the Apostle Paul's conversion Catherine was rapt in ecstasy and her spirit ascended so high that for three days and nights she gave not the slightest sign of life. Those present believed her dead, or on the point of death. A few, however, who understood what was happening considered that she had been taken up by the Apostle into the third heaven. Time passed, and the ecstasy ended; but her spirit, drunk with the heavenly things it had seen, seemed so reluctant to return to the things of earth that she remained in a sort of daze, like a drunkard who is stupefied but not asleep.
>
> When asked about what she experienced during her mystical voyages she answered: Father, my soul saw and understood everything in the other world that to us is invisible: that is to say, the glory of the Saints and the pains of sinners. I have already told you: the memory cannot keep anything of it and words are not adequate to describe it; but as far as I can I will try to tell you about it. You can be certain, then, that my soul contemplated the Divine Essence; that is why I am now always so discontented with being in the prison of the body.

Gopi Krishna (1903-1984)

Gopi Krishna was a yogi mystic born in India who began practicing meditation at the age of seventeen. He wrote about his mystical experiences, which he attributed to the rising of kundalini energy, in his autobiography, *Living with Kundalini*, beginning with his first such experience.[2]

> Suddenly, with a roar like that of a waterfall, I felt a stream of liquid light entering my brain through the spinal cord. Entirely unprepared for such a development, I was completely taken by surprise; but regaining self-control instantaneously, I remained sitting in the same posture, keeping my mind on the point of concentration. The illumination grew brighter and brighter, the roaring louder, I experienced a rocking sensation and then felt myself slipping out of my body, entirely enveloped in a halo of light. It is impossible to describe the experience accurately. I felt the point of consciousness that was myself growing wider surrounded by waves of light. It grew wider and wider, spreading outward while the body, normally the immediate object of its perception, appeared to have receded into the distance until I became entirely unconscious of it. I was now all consciousness without any outline, without any idea of corporeal appendage, without any feeling or sensation coming from the senses, immersed in a sea of light simultaneously conscious and aware at every point, spread out, as it were, in all directions without any barrier or material obstruction. I was no longer myself, or to be more accurate, no longer as I knew myself to be, a small point of awareness confined to a body, but instead was a vast circle of consciousness in which the body was but a point, bathed in light and in a state of exultation and happiness impossible to describe.

Gopi Krishna goes on to describe how the serpentine energy (kundalini) transformed his life, creating many difficulties and wonders.

Perhaps the most concise description of the mystical state is provided by the Mandukya Upanishad. It says that below the waking state lie the states of dreaming sleep and deep dreamless sleep. But beyond these is the fourth state (*turiya*), the transcendental state, which is characterized by pure unitary consciousness and bliss (ananda)—invisible, otherworldly, incomprehensible, without qualities, indescribable, the unified soul in essence, peaceful, auspicious, and without duality.

Summary

There have been hundreds of accounts of mystical union described in the popular literature. The experiencers universally describe entering into a clear, exalted state of ecstasy and undifferentiated, limitless consciousness in which their little self or ego merges with the underlying spirit of the universe or God. The descriptions of this life-changing event are strikingly similar and attempts to label these as hallucinations or products of abnormal brain chemistry would have to come to terms with this fact, and the fact that the experiences are both incredibly powerful and transformative. Mystics claim to have experienced a state of consciousness that is more expansive, knowing, and unitary than normal waking consciousness. They are seemingly witness to the supremacy of Consciousness over mind and matter and have experienced creation as unfolding from Consciousness. This is in complete opposition to the materialistic worldview that mind and consciousness are byproducts of matter (brain). Unfortunately, the experiences of mystics are personal and cannot be shared directly with us so there will always be skeptics who will label such experiences as "anecdotal" or products of a deranged or abnormal mind. However, the mystical vision of reality is more consistent with the nonlocal ontology, which is now a part of modern science.

7
Remembering Past Lives

BEGINNING AT AGE TWO, James, a native of Dallas, had recurrent nightmares of being shot down and unable to escape the burning wreckage of his Corsair during the war. As memories of this previous life surfaced, he recounted to his parents many details about his former life, including such things as the name of the aircraft carrier he took off from, the name of a friend he flew with, the name of a Japanese ship he strafed, and the location of where he crashed. These details checked out with actual historical records. By age eleven, James was no longer bothered by remembrances of his past life and death.[1]

Stories of this type by children three to six years old are more common than might be expected. Many children have distinct recurring memories of life as a person who died before they were born. They remember many details about that person, such as names, places they lived, details about their former parents and siblings as well as details of their death. Sometimes the details given by the child are specific enough that their parents or an investigator can identify the deceased person and check out whether the information given by the child is accurate. Memories of this type often fade after the child reaches the age of ten. One example, taken from a case studied by Jim Tucker, which recently gained national attention in the news media, is that of Ryan Hammons.[2] Ryan reportedly began having vivid nightmares at age four about a previous life. The next year, Ryan told his mother he wanted to go home to Hollywood and related stories about meeting actress Rita Hayworth, taking trips to Paris, and dancing on Broadway. The boy said he had once lived on a street with the word "rock" in it. His mother said his stories were detailed and extensive, unlike something a child would make up. One day, when going through some old Hollywood picture books, Ryan immediately identified a picture of a minor actor, who later became a Hollywood agent named Marty Martyn. Ryan was certain that he was this man in his previous life. Ryan's mother, Cyndi Hammons, could not find any more information about the man, so for help she approached Dr. Jim Tucker, associate professor of psychiatry and neurobehavioral sciences at the University of Virginia.

Dr. Tucker has spent more than ten years studying the phenomenon of reincarnation memories in children. Tucker was able to confirm many details about Martyn's life that the boy had provided. Such things as the fact that Martyn had danced on Broadway, traveled to Paris, worked with Rita Hayworth, been married five times, had two sisters, one daughter, and many other details about Martyn's profession, and even his having lived on a street with the name "rock" in it (825 Roxbury Dr.). In all Tucker was able to confirm fifty-five details about Martyn's life out of the fifty-six provided by Ryan. The one apparent error that Ryan made was saying that he had died at age 61, when it was thought he had died at age fifty-nine. However, after examining old census records, Tucker discovered that Martyn had been born in 1903, not 1905, making him sixty-one when he died.[3]

Ian Stevenson began collecting cases of this type in 1961 and Tucker took over the work after Stevenson died in 2007. To date they have collected and documented 2500 cases of children who remember past lives. Stevenson summarized many of his studies in his book *Children Who Remember Previous Lives: A Question of Reincarnation*.[4] Stevenson did extensive case histories on over sixty children whose memories checked out with the details known about a deceased person. Stevenson and Tucker do not claim that such memories of living in a previous body are proof of reincarnation, but feel they do demand an explanation. Anyone studying the data with an open mind would have to admit that there is strong evidence for the survival of memories and therefore some aspects of personality following death.[5] Such memories could not be stored in the brain but are characterized as noncerebral memories.

Stevenson also did extensive research on birthmarks and birth defects in children linked by childhood memories or their mother's knowledge of injuries (usually fatal) suffered by a deceased person.[6] One example taken from Stevenson's 2,268-page two-volume monograph on this subject and highlighted by Kelly,[7] may illustrate such a case. An Indian boy named Hanumant was born with a large hypo-pigmented birthmark on his chest. Before his conception, his mother had seen the body of a man who had been murdered in her village by a shotgun blast to the chest. Later she had a dream that the baby she was carrying was the reincarnation of this man, and between the ages of three and five Hanumant spoke as if he was this man in his previous life. Stevenson was able to find a close similarity between the postmortem report of the shooting victim and the boy's birthmark. In Hanumant's case, the mother had a premonition that her child would be linked to this man; however, in other cases reported by Stevenson, the link between various birthmarks and birth defects was not made until after the child was born and began to talk as if he were a particular person in his previous life.

Adults remembering past lives

Not only children have memories of living in another body in the past; many adults have also reported such memories. For example, actress Shirley MacLaine provided vivid details of her past-life experiences in her book *Out on a Limb*.[8] Remembrance of past lives may occur spontaneously in some individuals but is more common in meditators and most common in people undergoing hypnotic regression. While undergoing past-life hypnotic regression, both healthy individuals and those suffering from phobias have reported experiencing images, sounds, and smells as if they were in a different body in a bygone era. Raymond Moody, Jr. M.D. did extensive research on past-life regressions and reported on his own nine past-life experiences and those of many of his patients in his book *Coming Back: a Psychiatrist Explores Past-Life Journeys*.[9]

As it turns out, almost anybody who can be deeply hypnotized can be regressed by the hypnotist and have experiences of past lives. The experiences are most often visual and take on a life of their own without intercession by the therapist. Subjects identify strongly with a particular individual and feel the emotions of that person during the regression. Often these feelings mirror problems faced by the subject in their present life. Finally, most of the subjects feel like a weight has been removed from their mind after they relive a particularly traumatic experience from a previous life.[10]

It has become an increasingly common practice for psychologists to hypnotically regress people who suffer from neurotic phobias. Often they remember an incident that occurred in a previous body and time. Once they relive the traumatic memory, they are normally cured of their neurosis.[11,12]

Reincarnation in various religious traditions

Reincarnation was virtually a universal belief in the past. It was a central doctrine of Egyptian, Greek, and Roman polytheism, and in Christian gnosticism, shamanism, Native American, and African folk religions.[13] In addition, the Celtics of Great Britain and the Vikings believed in reincarnation, as well as the Ismaili's, a Shiite sect of Islam.[14] Today, the monistic Eastern religions of Hinduism (Vedantism), Buddhism, Taoism, Zoroastrianism, Jainism, and Sikhism universally and wholeheartedly teach the doctrine of reincarnation. In addition, Judaism, Islam, and Christianity have deep ties to this doctrine. In Judaism, there has been a fundamental belief in reincarnation or *gilgul* that goes back thousands of years. In modern versions of the Jewish faith, this belief has been largely ignored; however, it lives on in the orthodox and Hasidic communities and is central to Kabbalistic or mystical Judaism. Sufism, the mystical version of Islam, also teaches that if the soul cannot attain union with God it will return to earth to continue its path toward perfection.

There is ample evidence to suggest that early Christians believed in reincarnation.[15, 16] Gnostic Christians believed that they would be reborn in a new body following death if they were unable to attain perfection in this life.[17] By some accounts, before the rein of the Emperor Constantine gnostic Christians outnumbered orthodox Christians. However, when Constantine decided to make Christianity the official religion of the Roman Empire, he favored the orthodox wing because they had an established hierarchy. He also felt that the concept of reincarnation was a threat to his empire, since his soldiers might be less inclined to die for the emperor if they believed they would have to return to earth again instead of being dispatched directly to heaven. Therefore, most of the early writings and gospels that referred to reincarnation appear to have been deleted from the New Testament in the fourth century during the reign of Constantine. In the sixth century, the Second Council of Constantinople officially declared that the belief in reincarnation was a heresy. Any teaching of this doctrine was thereafter brutally suppressed. A few references to reincarnation appear to have made it into the four accepted gospels of orthodox Christianity. For example, Jesus told his disciples that John the Baptist was the prophet Elijah.[18] And there are several other passages in which both Jesus and John the Baptist are thought by some to have previously been Elijah.[19] At that time, the concept of returning to earth in a new body was a commonly held belief, and Jesus taught that a spiritual change had to take place so that one can obtain liberation.

Today belief in reincarnation is common among the world's population; by some estimates the number of people who believe in reincarnation exceed the number who reject the idea or have no knowledge of it.[20] Recent surveys showed that 20 percent of Americans believed in reincarnation and 26 percent of Canadians, while for Europeans the number was close to 30 percent.[21] For most people in the West this belief is not a matter of doctrine or faith; therefore, one might conclude that in many believers it results from an intuitive feeling that their individual existence does not begin with their birth and end with their death.

Genius

The spontaneous appearance of extraordinary skills, ability, or genius in young children, often in contrast to the expectations of their family, may be attributed to reincarnation. For example, Isaac Newton was born almost exactly one year after the death of Galileo and his life's work appears to be a continuation of that of his predecessor. Mozart displayed prodigious musical ability from his earliest childhood. Already competent on keyboard and violin, he composed music from the age of five and while still a boy performed before European royalty. Mozart received only rudimentary training in music from his father whom he surpassed in musical composition at an early age. Throughout history, one can find many other examples of individuals who from a very early age had a predisposition, motivation, or ability and who made extraordinary contributions to society. If indeed the body is only temporary but the mind (soul) is permanent, then it is no mystery why knowledge, abilities, and inclinations might be carried from previous lives to a current life.

Summary

If you ask most scientists today whether there could be life after death, they would probably say no. Any suggestion that people possess minds or souls that survive death would be labeled "vitalism," a superstition that was discredited a century ago. Moreover, those scientists that do believe in life after death probably do so as an article of faith. In the West that probably means they are Christian, Jewish, or Muslim—none of which embrace the concept of reincarnation. Hence in the West there is a strong bias against the notion of reincarnation. It is no surprise that scientists willing to investigate these phenomena are subject to ridicule; and children expressing memories of past lives are told that these memories are false or imaginary. As a result, most of the data about children and adults who remember previous lives comes from Eastern cultures that accept reincarnation as a matter of faith. The scientists that have investigated such phenomena point out that because people recall living in a previous body, it is not conclusive proof of reincarnation. They point out that such persons could be getting their information from the collective unconscious or by ESP. However, in Part III of this book, we will see why the spiritual worldview requires the continuation of individual consciousness following bodily death until the unit consciousness merges with Cosmic Consciousness.

8
Out-of-body and Near-death Experiences

During normal or waking consciousness, people have a clear sense that they are "in" their body. This awareness is altered in what is called an out-of-body experience (OBE). Here a person has a vivid experience of leaving their physical body. Often they feel themselves floating above their body and are able to witness events from this unusual perspective. The OBE is also a salient feature of the near-death experience (NDE), but differs in that the OBE is usually not associated with a close brush with death. OBEs most commonly occur during anesthesia, while falling asleep, and during lucid dreams. Psychologists label the OBE a dissociative experience arising from different abnormal psychological and neurological factors—an altered state of consciousness like a dream or hallucination. Few neuroscientists today consider the OBE to be evidence for the ability of the mind to leave the body and gain information that would not be available locally through the sense organs. Investigators also report that many of the features of an OBE can be reproduced using artificial stimulation of the brain and by hallucinogenic drugs such as LSD. The consensus among neuroscientists is that OBEs are of physiologic origin, but is this correct? While some of these experiences may have a psychophysiological explanation, there are good reasons to believe that a paranormal explanation is needed for some OBEs.

Take for example the case of a young woman named Sarah, who while under anesthesia for a routine gall bladder operation suddenly experienced ventricular fibrillation. Her heart was restored to normal rhythm using a defibrillator, and following the procedure she related to the medical staff and her family her experience during the operation. She described floating above her body, and from that vantage she could clearly hear and see many things that occurred as the doctors and nurses frantically tried to restart her heart. These included the layout of the operating room, scribbles on the surgery schedule board, color of the sheets, the nurse's hairstyle, and even the fact that her anesthesiologist was wearing mismatched socks that day. Her perceptions were remarkably accurate despite the fact that Sarah had been blind from birth.[1]

Like Sarah's OBE, many people report observing events from a unique perspective while floating outside their body. H. Hart did an analysis of almost three hundred published accounts of OBEs in which the person reported observing events that they could not have witnessed using their ordinary senses, and the details in nearly one hundred of these cases were corroborated by a second party or by other means.[2]

Hart also investigated OBEs in which a second person at a distant location had a vision of or felt the presence of the person undergoing the OBE. Such cases are called "reciprocal apparitions." If the knowledge received by persons undergoing an OBE were strictly due to an ESP phenomenon known as remote viewing (a form of clairvoyance), then it would be difficult to explain how another person could share in the experience.

Beginning in 2001, Sam Parnia and colleagues have investigated claims of OBEs by persons undergoing cardiac arrest. These so-called AWARE studies (AWAreness during REsuscitation) have looked into the experiences of over 1500 cardiac-arrest survivors in order to determine whether people without a heartbeat or brain activity can have documentable OBEs and near-death experiences. In some of these studies, objects or placards were placed on shelves that could only be observed from above. So far, only about 2 percent of the individuals in the study have reported full awareness compatible with an OBE and none have identified the hidden image. However, most of the OBE patients described having clear visual experiences from a vantage point outside their body, something Parnia claims would be impossible when the brain is shut down during cardiac arrest.[3] It seems likely that if the OBE is real, eventually someone will report seeing the hidden image providing further evidence supporting the reality of this experience.

For adepts in yoga, the out-of-body experience is taken for granted as one of the siddhis (powers) that may be acquired through intensive meditation. The term often used to describe this supernormal power is "astral projection." It is believed that the subtle or psychic body leaves the physical body but is still tied to the physical body by a cord, much like a kite is attached to the ground by a string. The psychic body is then free to roam the cosmos according to the will of the yogi.[4] Similar to other powers obtained by spiritual practice, students of yoga are constantly reminded not to pursue them because they can lead one astray and be dangerous to their psycho-spiritual health. Today several institutions specialize in the study of OBEs and astral travel.[5]

Near-death experiences

The near-death experience or NDE is characterized by a lucid out-of-body experience along with feelings of peace and bliss. Similar to the OBE, there will usually be accurate visual and auditory experiences from a vantage point outside the body, often seeing their lifeless body from above and witnessing attempts to resuscitate it. Next, there is usually the experience of entering a dark tunnel with a brilliant light beyond. Often loved ones are seen in the tunnel. Once they enter the light, there is the feeling of being in the presence of a being that radiates infinite, unending, and unconditional love. Often these experiences are complimented by a life review in which the dying person witnesses or relives thousands of past events simultaneously in an altered, omni-view state of consciousness. They also report witnessing how their actions affected other people both positively and negatively. Finally, the person may feel that there is a line, which if crossed will lead to death. They either feel or are told that it is not their time to go and describe being reluctantly drawn back to their body, usually because of commitment to family or loved ones. Almost universally, the person will describe the NDE as a life-changing event. Their attitudes, beliefs, and outlook on life are permanently and dramatically changed. Even among persons who were previously atheists, there is a certainty that God exists and that there is life after death.

NDEs are normally associated with a close brush with death, such as cardiac arrest, but may also occur during anesthesia or be triggered by a strong expectation that they are going to die but in which there is actually no physical possibility of death. During the NDE, they claim to experience being in a nonphysical body that is completely healthy, pain free, weightless, and blissful. They report being fully conscious and have full memory, judgment, and imagination. The images they witness in the "disembodied" state are described as highly vivid, with heightened awareness and clarity—more real than normal waking consciousness.

The history of NDEs goes back thousands of years and they have been reported in many diverse cultures. One of the earliest accounts was that of Plato in Book Ten of the *Republic* in which he described the experiences of a soldier who was thought to be dead and had been placed on a funeral pyre. There have been many additional reports of NDEs since that time, and in modern times, the medical literature has numerous such reports. Estimates place the prevalence of NDEs at 10–20% of patients close to death.[6]

Raymond A. Moody, Jr. probably did more to popularize the NDE than any other researcher. In his bestselling book *Life after Life*, he chronicled the accounts of 150 survivors of a near-death experience and first coined the term NDE to describe these accounts. Moody concluded that the NDE was strong evidence for life after death.[7] More recently, Jeffrey Long in his book *Evidence of the Afterlife: The Science of Near-Death Experience* argued that there are nine observations that prove the existence of life after death.[8] These were generated through the study of hundreds of NDEs and the consistencies of the accounts that he compiled over the years. These arguments include: (1) it cannot be medically explained how people experience consciousness outside their body when they are clinically dead; (2) blind people experience visual perceptions during their NDE; (3) children give NDE details similar to adults, though they may have never been exposed to this concept; (4) "life-review" experiences that tend to reflect real events. These observations, along with the others, are the primary basis for Long's assertion that the NDE data proves that there is life after death.

One of the most compelling reasons for believing that the NDE is a real phenomenon associated with the dying process is the remarkable similarity of the accounts, regardless of age, nationality, religion, race, culture, and other demographics. No two experiences appear identical, but even among children one or more of the elements mentioned above are reported. Certainly, predisposed cultural and religious attitudes play a role in the experience, since for example, Christians are more likely to say that they felt or saw an image of Jesus while Hindus are more likely to report seeing Krishna.

The NDE clearly challenges the prevailing opinion of neuroscientists that all mental activity can be attributed to the brain. What is most difficult to explain is how there can be a continuation of consciousness and even enhancement of mental awareness during a time when the brain is shutting down or which in many cases has stopped functioning because of a stoppage of blood flow. Not only is there lucid consciousness and vivid memory but also fully structured thought processes and the same sense of self that exists in normal waking consciousness. Often the experience is described as so beautiful and transcendent in nature that words simply cannot describe it, and its effect on the person may be felt for decades.

EEG studies of people suffering cardiac arrest indicate the absence of gamma waves, normally associated with waking consciousness, and within a few seconds of circulatory collapse the EEG will display a flat line, which is one of the characteristics of death (the others being lack of heart beat, respiration, and brainstem reflexes).[9] As mentioned before, Sam Parnia, a cardiologist who has done extensive research on the resuscitation of patients suffering cardiac arrest, has concluded that the oxygen-deprived brain of such a person could not possibly produce the images and lucid consciousness that is a hallmark of the NDE. In fact, the ordinary unconsciousness that accompanies such events is typically associated with confusion and impairment of memory (amnesia). It is also very difficult to explain how persons undergoing such insults to the brain could obtain and later relate accurate and verifiable information about events that took place while they were unconscious and in some cases considered clinically dead.

Prevailing physiological explanations for the NDE

Perhaps the most popular explanation of the NDE put forth by neuroscientists in the brain-equals-mind camp is the theory that during the process of dying, as brain shuts down, people may experience a dream-like state and/or hallucinations as the subconscious mind takes over. However, this theory fails to adequately explain how people who are pronounced clinically dead could have such lucid consciousness, a consciousness that is described as more vivid than normal day-to-day consciousness—nor does it explain many of the other elements of the full-blown NDE.

Some neuroscientists have objected to the notion that a flat-line EEG is synonymous with a lack of brain function. There is ample evidence to suggest that there can be deep-seated neuronal activity in the brain that does not show up on a scalp EEG. Hence, undetected brain activity could be going on and be responsible for the NDE. The problem with this theory is not that there might be brain activity, but if there could be the type of brain activity that is considered necessary by contemporary neuroscience as a requirement for conscious experience. The answer appears to be no, since the hypothesis that there is a brain basis for consciousness depends on the ability of widely separate regions of the brain to coordinate with one another—something that could not occur under these circumstances.[10]

Another theory is that the experiences do not occur when they seem to occur but either during the initial shutting down of the brain due to a lack of oxygen or right afterward as the brain is being restored to normal functioning. One problem with this hypothesis is that such insults to the brain are known to be associated with confusion and memory loss. Secondly, many NDEs have "time stamps" in which the person reports specific events during the time their heart was stopped. These include witnessing resuscitation attempts and frantic attempts to call for help.

Skeptics of the paranormal explanation of the NDE have argued that many of the features of an NDE can be reproduced by drugs or other means. For example, the massive release of endorphins and enkephalins (endogenous opioids) in the brain might cause blissful feelings and an altered state of consciousness. Others have pointed out that the administration of ketamine (an anesthetic) is capable of reproducing many of the features of the NDE. There are also similarities between NDEs and the loss of consciousness due to high G-forces, such as those induced in a centrifuge for flight training. These include tunnel vision, lights, and floating sensations. Others have pointed out the similarity between NDEs and hallucinations caused by psychedelic drugs such as LSD and DMT. The prevailing attitude of neuroscientists that are skeptical of the idea that the NDE represents a paranormal phenomenon is that every element of a near-death experience can be replicated with drugs, with anoxia (lack of oxygen), with lack of blood flow, or by turning off circuits in the brain.

The biggest challenges facing such skeptics is how these experiences can seem so real and memorable and have such profound aftereffects if they are simply hallucinations, dreams, confabulations, or drug-induced brain effects, when none of the other proposed physiological explanations have these effects.

Another challenge to the brain-based explanation of the NDE comes from Raymond Moody. He points out that elements that we think of as a near-death experience—leaving one's body, going into the light, seeing a panoramic review of one's life, seeing deceased relatives—occur not just to people who have the NDE and are brought back, but to healthy and uninjured bystanders at the bedside of the victim. These so-called "shared death experiences" are not uncommon, and the bystanders say that they also left their bodies and accompanied their dying loved ones partway toward the light, or that they saw the apparitions of relatives and friends of the dying person come into the room. Moody has accounts of hundreds of people with shared death experiences—some of whom had the identical experience to the person with a NDE. They reported experiences characteristic of the NDE but they themselves were perfectly healthy.[11] If the cause of these near-death experiences were a physiological insult to the brain, how could bystanders who were not ill or injured have essentially the identical experience?

Summary

The popular current hypothesis of the neurosciences is called epiphenomenalism, which in effect says that there is no independent reality to what people experience as consciousness. It is a secondary byproduct of the primary reality, which is the brain and the electrochemical events going on in the brain. Hence, most neuroscientists freely dismiss the OBE and NDE as aberrant brain conditions. Although neuroscience has yet to propose a meaningful model for how consciousness and self-awareness can arise from strictly neurochemical reactions in the brain, it would not be fair to underestimate the great progress made by neuroscience in the last few decades in the understanding of the brain and how it functions, and the indispensable role it plays in the perception of reality. Researchers in the field of neurophysiology have uncovered strong evidence that almost all mental states and conditions have an accompanying brain-based cause, i.e. a physiologic etiology. However, just because mind is dependent on brain for its function does not mean that brain is *its cause*.

If one accepts the prevailing doctrine that matter is the fundamental building block of reality, then it is natural to overlook the distinction between cause and effect and dismiss all the evidence to the contrary as anecdotal, superstition, bad science, or simply impossible. The history of science has many examples in which the large majority of scientists favored the status quo even in the face of anomalous evidence that in hindsight should have offered clear indications that the prevailing theory was incorrect.

The data on OBEs and NDEs indicates that these experiences involve altered states of awareness and vivid and unforgettable images and sensations—mental functions that could not possibly occur according to current neurophysiologic models of how the brain produces mind. The alternative hypothesis that mind is not body and may separate from it under certain conditions, including death, appears to be a more plausible explanation for these experiences.

9
Extrasensory Perception

EXTRASENSORY PERCEPTION OR ESP consists of several related phenomena. These are clairvoyance, telepathy, precognition, and psychokinesis. Clairvoyance (remote viewing) is the ability to obtain information about places, things, or events at a remote location. Telepathy is the transfer of thoughts or feelings between individuals at a distance without any apparent physical means. Precognition involves perceiving information about future events before they occur, and psychokinesis is the ability of the mind to influence matter, space-time, or energy. These phenomena are also known as psi phenomena and research into them is known as the field of parapsychology.

If ESP phenomena exist, it would be a huge challenge to the materialistic worldview. This becomes obvious if one considers the fact that the existence of ESP would negate the scientific, naturalistic, reductionist, materialistic, physicalist, atheist doctrine that consciousness is a product of brain and that nonlocal interactions of consciousness are therefore impossible. Hence, it is not surprising that this group comes down hard on ESP research and has attempted to disparage the data supporting it. Take for example the following statement from Sean Carrol, physics professor at the California Institute of Technology and self-proclaimed atheist:

> The only problem is, parapsychology is not science. It's pseudoscience. From a completely blank-slate perspective, one can certainly pose scientific questions about whether the human mind can tell the future or read minds or move objects around without touching them. The thing is, we know the answer: no. The possibilities have been investigated and found wanting; more straightforwardly, they would violate the known laws of physics. Alchemy was science once, but it's not any more. Not all hypotheses are equally worthy

of our respect and attention; sometimes we learn that a particular idea doesn't work, and move on with our lives.[1]

We might wonder what laws of physics this physicist believes ESP violates. Perhaps it is the law of gravity, because the mind sometimes floats out of the body. Perhaps it is the law of electromagnetism because quantum particles and minds are believed to communicate faster than light speed. What this scientist seems to be saying is, in my opinion, this is not how nature *ought* to work. Carrol had more to say about parapsychology in a recent debate on whether there is life after death:

> Even if we put plausibility aside, a hundred years of parapsychological research has not been able to produce one bit of compelling replicable evidence that shows that it's a real phenomenon.[2]

Here he parrots the standard argument against ESP by those who are wedded to materialism and its bottom-up worldview. Nonetheless, the fact is, skeptics need to deal with ESP since no amount of dismissive hand waving is going to negate the thousands of well-controlled scientific studies that have provided incontrovertible evidence for its existence. We will outline the evidence in the next few sections.

Telepathy

The word telepathy actually means "feeling at a distance." However, since it was coined by renowned psychologist F. W. H. Myers in 1882, it has come to mean communication between two minds. Experimental studies of telepathy have been going on for at least 125 years and it is safe to say that if no positive data were obtained studies of this type would have been discontinued long ago.

Probably the best example of the modern research method for telepathy is the ganzfeld method. It is well established that a quiet, peaceful environment with limited sensory inputs is optimal for receiving information from another mind. Hence, the receiver in such experiments is situated in a comfortable reclining chair with halved ping-pong balls placed over their eyes and white noise playing on headphones. A soft red light illuminates their face. These conditions create a restful state similar to one produced in a sensory deprivation chamber or floatation tank. Next, an assistant is asked to select one target pack of four images from a pool of many such packets. Then one target image is selected from the four images in this packet, which is in an opaque envelope. This envelope is given to the sender, who is in a distant, isolated room called the sending chamber. The sender unseals the envelope and mentally tries to send the image to the receiver over the next fifteen to twenty minutes, after which the receiver is returned to a normal environment and asked to select the image that was "sent" from the four images that were in the original packet. Variations to this procedure include using video clips and automating the selection process using a computer. The sender is asked to rank the images from one to four. If the actual image (or video clip) is ranked first, it is considered a hit, which is expected to occur by chance one fourth of the time.

From 1974 to 1997, there were over 2500 ganzfeld sessions reported in forty publications by researchers around the world. In a 1985 meta-analysis[3] of the published studies that provided hit data, twenty-three out of twenty-eight studies resulted in hit rates greater than chance with odds against chance of a billion to one.[4] After receiving some criticism from skeptics, a new round of experiments were started in 1983 by Honorton and colleagues that were computer-controlled. During the six years of the study using 354 volunteers, a 34 percent hit rate was achieved.[5] These results were similar to the nonautomated experiments indicating that people can sometimes receive information at a distance without the use of the five senses.

There have been numerous, well-documented reports of telepathy between identical twins—some separated by birth and later reunited. The only reasonable explanation for the uncanny connection that clearly exists between some twins is that their minds are entangled, just as their bodies were in the womb.[6]

Clairvoyance

The only difference between clairvoyance and telepathy is that no sender is required. Information is obtained about something from a distance, and thus clairvoyance is also called "remote viewing." There have been many anecdotal reports of clairvoyance involving a dream, sudden knowledge, or a strong feeling that a loved one was injured or had died. The time and nature of the perceived incident often coincides with those of the accident, well before they receive news of the event.

The first scientific studies of clairvoyance began in the 1880's using targets such as playing cards. Over the next 125 years, the experimental techniques have been refined in order to better control the experiments, and five-symbol card decks have become standard. The cards may be in sealed envelopes, behind opaque screens, separated by distance, or even displaced in time. The hit rate using all these methods in a subset of tightly controlled card experiments run by twenty-four different investigators from 1934 to 1939 in over nine hundred thousand trials was significantly above the 20 percent expected by chance.[7] J. G. Pratt calculated that if all the clairvoyance card tests conducted from 1882 to 1939 were combined, the odds against chance would be more than a billion trillion to one. This was a four million trial database performed by sixty investigators in many different countries and reported in nearly two hundred published reports.[8]

Precognition

The various forms of ESP or psi phenomena are very difficult to distinguish and probably overlap in such a way that distinctions have no real meaning. For example, how can one know in telepathy experiments such as the ganzfeld method that the receiver is not getting their information clairvoyantly? Similarly, the clairvoyant experience that someone very close has been injured or died may be a result of receiving information about a future event—the time when they learn about the accident by normal means. This would be labeled precognition. Finally, suppose an experiment is set up where the test subject is asked to predict the throw of a die or flip of a coin. How would it be known whether a positive result was actually precognition rather than psychokinesis? It turns out that precognition could explain many psi phenomena. It would simply mean that one knew something about the future that was later verified when the event was actually experienced. Such an idea is not that outlandish when it is considered that physics demands that the future is already "out there" in the realm of unchanging space-time.

There are innumerable examples of people who have accurately prophesized their own imminent death, a disaster, or other future event. Various cultures and religions take prophesy quite seriously and it forms the basis for many mythologies. Even today, while it is common to dismiss many of the stories of lore as myths and superstitions, it is such a common experience for people to have an intuitive feeling about something that later happens, that prophesy and fortune telling are accepted by a majority of people worldwide. This does not mean precognition is true, but it does suggest that it may not be that rare for a person's nonlocal mind to bump into their future.

However, the experimental "proof" of precognition comes from thousands of well-controlled scientific experiments. The typical experimental protocol is to have a person guess in advance which target will be displayed from a randomly selected group of possible targets. If the correct target is selected, then this is counted as a hit. Many studies of this type have been conducted since 1935, and in almost all of the studies, the null hypothesis (that results are due to chance) was rejected. Combining all the results from 309 studies, 113 articles, sixty-two different investigators, and almost two million individual trials gave odds against chance of 10^{25} to one.[9] Odds of even ten thousand to one indicate that chance occurrence must be eliminated as an explanation for the results.

Experiments also show that unconscious autonomic responses to stress occur before the stressful situation is witnessed consciously. The technical term for such a response is "presentiment." It can be measured by attaching sensors such as skin conductance, heart rate, or blood flow in the fingertip, and then having a computer display a series of photos, some of which are calming and some of which are disturbing. Naturally, the monitors show a strong response immediately after the subject sees a disturbing photo and no reaction following a calm photo, such as a field of flowers. The surprising thing is that the computer registers a significantly larger fright-and-flight response just before the emotionally charged photo is shown than before the calm photo. The opposite effect of a dropping heart rate can also be seen just before the calm photos are shown.[10] Experiments like this in which an unconscious autonomic response precedes the actual insult have been replicated and indicate that the nervous system is capable of anticipating future events.

Psychokinesis

Does mind affect matter? In one sense, the answer is easily demonstrated. Think about raising the arm and it is raised. The definition of mind is that it is nonmaterial (i.e. your mind is separate from your brain) and it moves matter (body) all the time. The effect of mind over matter is what is meant by psychokinesis, but body movement is not the type of psychokinesis that is of interest here. The most likely explanation of how mental intention affects the body (matter) is that it collapses the quantum wave function of the brain in a certain way, creating the neuronal activity needed to move the arm. More interesting from the standpoint of psi research is whether mental intention can move or influence something in the physical world that is separate from the body.

It seems that most people believe that they have some influence over physical objects or events. Most gamblers attempt to influence the throw of dice in a game of craps or the spin of a slot machine. Nearly all golfers, including top professionals, "talk" to their ball, hoping it will listen and behave a certain way after they hit it. Belief that intention can somehow affect the world or bring good luck is pervasive. The question is whether this is possible or just a myth. The scientific method seems to provide the answer.

Experiments studying the effects of mental intention on the throw of dice have been going on for almost eighty years and provide conclusive evidence that such effects are real. A meta-analysis of 148 independent experiments indicated that the odds that mental intention does not influence dice are statistically one in a billion.[11] Interestingly, when no intention is applied to the throw of dice, researchers found that the hit rate was exactly that predicted by chance. Do these results prove that mind can affect the throw of dice? One would have to conclude, based on the high quality of the scientific controls employed and the reproducibility of experiments by more than fifty investigators, that the answer is yes.

Today most of the studies investigating psychokinesis involve experiments trying to affect random-number generators (RNGs).[12] Such experiments are easier to control and can involve trillions of randomly generated bits (ones and zeros) of information, which can be recorded automatically and with ease—in fact, the whole test can be automated eliminating operator bias or error. The effect also tests mind-matter interaction on a very small or quantum scale, which appears to be how mind fundamentally affects matter. Radin and coworkers performed a meta-analysis of 152 published reports from 1959 to 1987 describing 597 experimental studies in which persons were instructed to try to have the RNG generate an excess of either zeros or ones; and another 235 control studies in which no mental intention was used. Overall, the experimental studies generated a 51% hit rate while the controls were exactly 50%, as expected. Odds that such results would occur by chance are one in a trillion.[13]

The data on how mind affects RNGs has also shown that being close to the RNG has no significant advantage over being at a distance. Moreover, the intention of two or more people generally has a greater effect than a single individual. In addition, time is irrelevant to the results. For example, if a person's intention is to have a random-number generator write more zeros than ones on a floppy disk, it does not matter whether this intention is applied before, during, or after the disk is written. There will generally be the same statistically significant excess of zeros over ones in all three instances.

Recently, the massive data set from the Princeton Engineering Anomalies Research Laboratory (PEAR) on studies of human intention affecting RNGs has provided very strong confirmation that such mind-matter interactions do in fact exist. The PEAR program, which has been going on for nearly thirty years, is studying the interaction of human consciousness with sensitive physical devices, systems, and processes. In an analysis of the published PEAR data collected over the course of seven years from a group of subjects attempting to influence random-number generators across millions of trials, the observed effects were small (about one tenth of one percent), but over the databases, they compounded to statistically significant deviations from chance behavior (four thousand to one).[14]

Another study of this type dubbed the Global Consciousness Project was inspired and is currently being directed by Roger Nelson at Princeton. Begun in 1998, it utilizes a worldwide network of RNGs that collect randomly generated bits along with their time stamp and send the data to a server at Princeton. Data from this ongoing study show that there is a significant spike in nonrandomness coinciding with major global events, such as the death of Princess Diana, Y2K, 9/11, and the funeral of Pope John Paul II.[15] Numerous other globally important events caused significant spikes in the network of RNGs, indicating that the collective consciousness of humans on planet Earth have a small but reproducible effect on matter.

Answering the critics

Naturally, results like those outlined above pose a real and perhaps an insurmountable challenge to the materialistic worldview, which posits that it is impossible for mind to perceive or affect anything outside the body. Therefore, it is no surprise that parapsychology and research on psi phenomena have received scathing criticism from skeptics who are wed to materialistic dogma and who are quick to call all such studies "pseudoscience." In general, scientists who do such research are applying their time and talent to this work not because of any preconceived idea about reality but because of a real scientific curiosity; and they are not immune to the criticisms from those who think all such investigations are phony. Materialists are convinced that there has to be another explanation—after all, psi is impossible. The following is a list of criticisms and how scientists investigating ESP have answered these criticisms:

- *The studies are not well controlled.* If better controls were used then only chance results would be obtained. Actually, the data is contrary to this. As studies were increasingly controlled and refined in face of criticisms, the results did not become less significant but actually increased in statistical significance vs. random chance.[16] In addition, some researchers, being sensitive to such criticisms consulted with professional mentalists and magicians who could find no flaws in their methodology.
- *The studies are not reproducible.* Actually, the studies are reproducible. Many studies have been independently replicated. For example, presentiment studies were replicated by an independent investigator using his own equipment.[17] Moreover, there have been hundreds of ganzfeld telepathy experiments, all getting positive results.
- *Nonconforming data is not reported.* This is the so-called "waste basket" hypothesis that only studies showing positive results are published, which creates bias in the data. If the unpublished negative results were included in the data then the meta-analysis would not show positive results. The problem is the statistics do not support this criticism. For example, there would have to be fifteen unreported trials for each reported ganzfeld study. For the ESP card experiments it would take about thirty thousand unsuccessful studies out of thirty-four reported studies to reduce the odds to chance.[18] Hence, considering the time and effort it takes to conduct even one study, selective reporting cannot possibly explain the positive results obtained for telepathy, clairvoyance, precognition, and psychokinesis.
- *Faked or fraudulent results.* Investigators really want to believe that psi is real and either consciously or unconsciously manipulate the data in

order to achieve their expectations. While there have been a few cases of fraud or error, this is no reason to dismiss the vast majority of data. And while it is true that skeptics often get poorer or negative results when they attempt to replicate experiments of scientists who study psi phenomena, accusations that seasoned scientists, most of whom work at well-respected centers of higher education, would purposefully misreport their findings is patently ridiculous. Such scientists are fully aware of the controversial nature of their research and have every reason to want to protect their professional reputation. Therefore, they take a much more rigorous approach to insure that their work is of the highest standard possible—much more so than typical psychological studies. Furthermore, even if one or two studies out of two hundred were biased or flawed, it would not explain the other 198 studies, which gave positive results. In addition, statistical outliers are typically not included in a data set undergoing a meta-analysis.

- *Science is constantly explaining the supernatural in natural terms.* Throughout history, science has discovered that many phenomena that were earlier thought to be paranormal are actually explicable in natural terms. Of course, this is a phony argument since it says nothing about whether ESP is real, only that other phenomena, originally thought to be mysterious, were later explained by science.
- *Psi effects are tiny, irreproducible, and can only be shown to occur using statistics.* This is an invalid criticism since the same can be said of most studies in the "soft" sciences. In addition, results are reproducible and in some studies, e.g. using intention imprinted devices (see below), the effect is so robust that statistics are not required to demonstrate the scientific validity of the effect.
- *There is no theory to explain psi.* This is simply untrue. As we will see in Part III of this book, psi phenomena can be simply explained by the existence of unconscious mind, which is a collective mind, nonlocal in nature, that connects to the wholeness, which is at the very foundation of reality.

Other psi-related phenomena

Probably everyone is familiar with a friend or relative that seems to have particular problems running equipment or electronic devices such as computers. Sometimes all that person has to do is touch the machine and it will stop working. Sensitive electronic equipment seems particularly sensitive to such "gremlins," and persons with a reputation for "damaging" equipment or ruining experiments are sometimes asked to stay away from the lab.[19] On the other hand, some people have a way of fixing and getting machines and equipment to function smoothly. Can human expectation or intention influence equipment or is this simply superstition based on coincidence and selective memory? Data collected by PEAR clearly shows that such effects are real. In these studies human operators attempted to bias the output of a variety of mechanical, electronic, optical, acoustical, and fluid devices to conform to predetermined intentions, without recourse to any known physical influences. Absent intention, all of these sophisticated machines produced strictly random data, but when human intention was used, the experimental results deviated from chance in such a way that it could only be attributed to the conscious intent of their human operators.[20]

Over the course of almost thirty years and millions of trials studying human-machine anomalies, PEAR scientists concluded that intentions, emotions, and attitudes of the human operators do affect the sophisticated equipment that is used to generate data for modern science. Despite the small scale of the effect, this needs to be taken into account in future experiments utilizing such equipment.

Another line of mind-over-matter interactions comes from studies of Intention Imprinted Electrical Devices (IIED). In this experiment, a specific intention is imprinted into a simple electronic device (usually by an experienced meditator) with the intent to affect a measurable change in some experimental condition. For example, the pH of water can be increased or decreased significantly (0.5-1.0 units) and reproducibly by placing an activated IIED near the water as it is being monitored with a pH meter. The pH responds in exactly the way the device was intended, and the pH does not change when the IIED has no intention imprinted on it.[21] These same investigators also found that such things as the UV spectrum of DNA-water solutions, the thermodynamic activity of enzymes, the growth rate of fruit-fly larva, the temperature and conductivity of water, and the breakdown voltage of a gas-discharge tube[22] were altered by an imprinted device. All these effects are reproducible in the direction intended and robust to the point that no statistical analysis of the data is required to demonstrate that the effects are real.[23]

Several experiments have shown correlations between the electroencephalograph (EEG) traces of paired individuals. In such experiments, one person (the sender) is subject to a regular stimulus such as a flashing light, and the EEG of the distant receiver is observed to see if it jumps at the same rate as the brain of the sender. Results indicate that there is a significant and reproducible correlation for paired individuals and no correlation when there is no pairing.[24]

Carefully controlled scientific studies of mediums conducted by Dr. Gary Schwartz at the University of Arizona strongly suggest that they can obtain accurate information about people, places, and events.[25] The mediums normally attribute this to communication with discarnate entities, but they could also be obtaining the information by ESP. Schwartz also placed a semiconductor charged-coupled device (CCD) in a light-tight box in a dark room and recorded luminous discharges that he attributed to luminous beings. However, this could also be explained as a man-machine anomaly in response to human consciousness rather than that of a bodiless mind or being.

During the 1970's the US Department of Defense funded research on remote viewing to determine whether it could be used for espionage. The program began at the Stanford Research Institute (SRI) and in 1990 moved to the Science Applications International Corporation (SAIC), a defense contractor. The program ended in 1994 after twenty-four years and twenty million dollars of US government funding. In such experiments, a talented or gifted viewer is asked to sketch or describe a target facility. In some studies, a sender visits the site and attempts to send mental images of the site to the viewer telepathically. For testing purposes, a secret facility in the US might be used. For espionage purposes, the viewer might be asked to describe a hidden facility in a foreign country in order to enhance and corroborate other intelligence about it. In 1976 SRI researchers Puthoff and Targ published their positive results using primarily one highly gifted viewer (Ingo Swann).[26] Their work has been criticized by skeptics for lacking sufficient controls. However, in 1988 May and colleagues analyzed all the psi experiments conducted at SRI over a sixteen-year period, most of which were remote viewing tests. The data set consisted of over 150 experiments and twenty-six thousand trials. The statistical analysis indicated odds against chance of 10^{20} to one.[27] It is not known whether the CIA or other US intelligence agencies currently utilize remote viewing as part of their intelligence-gathering procedures. Nor is it known whether Russia or any other country currently utilizes psi to gather intelligence, but it is known that beginning in the 1960's the Soviet Union had an active program investigating psychic phenomena.[28]

Psychic Savants

Generally, ordinary people possess no more than a tiny capability for ESP and it takes a large number of people and statistical analysis of the data to demonstrate an effect that cannot be explained by chance alone. Occasionally a person will come along that possesses extraordinary psychic abilities—a so-called psychic savant. Such individuals are rare, but their ESP abilities defy all explanations purveyed by materialists. Perhaps the best known example of such a person was Edgar Cayce.

Edgar Cayce was born in Kentucky in 1877 and received only an eighth-grade education before he was called to work on the family farm. At the age of twenty-three, after suffering from chronic laryngitis, Cayce sought help for his condition from a hypnotist. He was told to perform self-hypnosis in order to obtain a permanent cure. After entering a self-induced trance, he described in detail the ailment and its cure. This was his first psychic reading and over fourteen thousand were to follow before his death in 1945. He performed his readings while in a trance and after awakening had no recollection of what he said. Most of his readings were for people seeking medical diagnoses and treatments. All he required was the individual's name and location. He would give an accurate description of the illness or physical problem as though he had X-ray vision and then offer a treatment for the condition. The accuracy of his readings was astounding and the terminology he used was not that of an unschooled individual but of someone highly trained in medicine.

Cayce also did readings on geology, chemistry, physics, electricity, history, political science, sociology, and anthropology. Interestingly, he confirmed the validity of reincarnation and indicated that some medical conditions and birth defects could be traced to events occurring in previous incarnations (i.e. karma). Cayce also made numerous prophecies that turned out to be uncannily accurate, such as the stock market crash of 1929 and the start and combatants of World War II.[29] A possible explanation for Edgar Cayce's psychic abilities is that while in a trance he tapped into the collective unconscious or Cosmic Mind. We will discuss this more specifically in Part III of this book.

History is replete with stories of people sporting superpowers or yogis demonstrating siddhis such as levitation. Some of these individuals, for example Jesus, are thought to have been born with these abilities while others (e.g. Buddha) developed their extraordinary abilities through intense spiritual practices—such as yogic exercises and meditation. There is ample scientific evidence to support the hypothesis that such supernormal or psychic powers exist in a few special individuals.[30]

ESP in animals

Many species of animals exhibit homing behavior that seems to defy rational explanation. Examples include salmon, many species of birds, and some insects, such as monarch butterflies. Scientists have posited complex explanations for such abilities, such as magnetic navigation, sun navigation, and sense of smell or a combination of sensory cues. However, the magnetic field of Earth changes constantly over time and its position changes over the surface of the planet. It is also affected by sunspots, iron deposits, and magnetic anomalies and is so "noisy" that it is unlikely that it could be used to provide anything but a general direction to a homing animal. Sun navigation depends on both surface position and time. Animals would have to have an extremely accurate internal clock to utilize this form of navigation. Mariners can obtain their latitude using a sextant and an accurate watch, but longitude is not easily measured using the sun. Therefore, navigation based on the sun can at best provide homing or migrating animals with a general direction—not precise location.

A keen sense of smell has been offered as a possible explanation for a homing instinct in animals such as salmon and dogs. Yet, it is unlikely that a particular stream would maintain the same smell signature between the time the salmon were hatched and when they returned to their spawning beds. Similarly, there have been many reported cases where a dog found its way to its owners at a new location weeks and even months after it was lost.

One such example is related by Larry Dossey. Bobbie, a female collie, was traveling with her family from Ohio to their new home in Oregon. During a rest stop in Indiana, Bobbie ran off and could not be found. After many hours of searching, the family gave up the search for Bobbie and continued to Oregon. After about three months, Bobbie turned up at the doorstep of their new home in Oregon. Bobbie had never been to Oregon and there was no mistaking her.[31] Many other stories of this kind have appeared in scientific literature and popular media.

The homing ability of some animals is a confounding mystery to science. One can postulate a sensory explanation for such behavior, including the utilization of several senses simultaneously, but the simple explanation for such behavior and ability is that the animals are obtaining nonlocal information in a manner similar to human ESP.

Throughout history, there have been reports of animals' "sixth sense" in detecting hurricanes, earthquakes, tsunamis, and volcanic eruptions before the event happens. Examples include dogs and cats and other animals acting strangely, birds taking flight, and rats leaving buildings before the ground begins to shake. This mysterious ability may allow some animals to sense geophysical changes in the earth before they happen.

For example, it was reported that before the giant tsunami hit the coasts of Sri Lanka and India on December 26, 2004, elephants screamed and ran for higher ground, dogs refused to go outdoors, zoo animals rushed into their shelters and could not be enticed to come back out, and flamingos flew to higher ground.[32] At the hard-hit Yala National Park in Sri Lanka, stunned wildlife officials reported that hundreds of elephants, leopards, tigers, wild boar, deer, water buffalo, monkeys, and smaller mammals and reptiles had escaped unscathed.[33] Rescue workers found surprisingly few dead animals at the park.

There has been at least one example where authorities successfully forecast a major earthquake, based primarily on the observation of the strange behavior of animals. In February of 1975 a 7.3 magnitude earthquake hit Haicheng, China. Haicheng had approximately one million residents at the time of the earthquake. Chinese officials ordered that the city of Haicheng be evacuated about twelve hours before the earthquake struck, believing that there was a high probability of an earthquake occurring. The prediction was based on widespread reports of unusual animal behavior. Only a small portion of the population was hurt or killed because of the evacuation. It was estimated that the number of fatalities and injuries could have exceeded 150,000 if no evacuation warning had been given.[34]

Some animals seem to be able to sense that a person is about to die or have a medical emergency. A recent example of this was reported in the July 2007 issue of the *New England Journal of Medicine* about a cat named Oscar that seemed to be able to predict the deaths of patients in a nursing home in Providence, Rhode Island. Just before patients died, Oscar would sit down by their beds and would become very upset if forced out of the room before the patient died. The article sited at least twenty-five successful predictions of impending death by the cat.

Dogs have been trained to give warning of an impending epileptic seizure in people prone to this condition. They are specifically trained to give warning to their owners so that they can take appropriate precautions before the seizure strikes. Whether the dog smells something from the person that provides a clue to an impending seizure or just uses its "sixth sense" is unknown. Dogs are also known to be able to detect cancers in people, even at an early stage. It has been postulated that it is their keen sense of smell that allows them to detect malignancies in humans, but this is only speculation. Another explanation for the uncanny ability of animals to sense things beyond the purview of the five standard sense organs is that they tap into nonlocal mind.

Why is the scientific evidence for ESP being ignored?

Anyone who studies with an open mind the massive and incontrovertible evidence supporting the existence of ESP in humans would have to admit that it is a scientific truth. Given this fact, one might wonder why this truth has not filtered down to most scientists or to the public. There may be a number of reasons for this as outlined below.

- *Denial.* Probably the most common reason for not considering the evidence in support of ESP is simple denial. It cannot be real, since if it were real it would overthrow the materialistic worldview. There is simply no way mind could operate nonlocally or affect an object if it were solely a product of brain. Such denial is based on a preconceived opinion of how nature ought to work and is similar to the dogmatic insistence that, for example, evolution is false because it is contrary to the biblical story of creation. The skeptics have made up their mind and therefore there is no reason to consider evidence to the contrary. A somewhat more scientific argument is that something nonphysical (mind) could not affect something physical (brain). Such an interaction would require the exchange of energy, which would violate the law of conservation of energy—ergo mind cannot exist separate from the brain and thus ESP is false. Actually, science says the opposite is true. The brain is a quantum mechanical system and nothing physical could cause the wave function of the brain to collapse because it would be part of the system. Without such collapse, not a single neuron could fire. Observation of the quantum mechanical brain by mind (or consciousness) is a necessary requirement for it to leave the realm of potentia (possibilities) and the domain of the wave function, and enter the realm of physical reality—no exchange of energy is needed.

☐ *Metaphysics has no place in science.* Historically, religious beliefs often clashed with science. Scientific discoveries were sometimes dismissed by powerful religious institutions when the scientific findings contradicted church doctrine. For example, Galileo was forced by the Catholic Church to renounce his findings that Earth revolved around the sun or face imprisonment. Because of the tension between science and religion, science had to disconnect itself from anything religious or spiritual in order to proceed with its task of describing reality in a nonsuperstitious, nonmetaphysical, and fact-based manner. The spirit behind this movement in science has been mostly positive, but it has caused science to swing to the other extreme, physicalism (aka scientism, materialism, reductionism, etc.), which essentially denies that there can be an alternate, nonmaterial-based explanation for phenomena that do not fit the reductionist doctrine. As a result, attempts to study or explain phenomena in nonphysical terms have been openly suppressed, ruthlessly criticized, or simply dismissed as pseudoscience by many of today's scientists.

☐ *Specialization in science makes it harder for people to consider an alternative to materialism.* Science today is incredibly specialized. In order to be at the forefront of scientific research one must normally focus their energy on a very narrow slice of the pie. As a result, most scientists today have little inclination to study what might be considered philosophical questions about the nature of reality (metaphysics). It is natural to go with the majority opinion that matter is the ground substance from which reality springs. To suggest otherwise leaves one open to scorn and possible harm to their professional reputation.

☐ *The public is more open to the idea that psychic phenomena are real but still tend to follow the lead of scientists.* In one survey the percentage of people that believed that psychic phenomena were real was 68 percent.[35] This is considerably higher than the percentage of scientists who believed (only 6 percent in the same survey of members of the National Academy of Sciences). We might suspect that the scientists are simply better informed than the public about ESP. However, there is no evidence to support this. More likely, they are simply more biased and less open to new ideas, since they are devoted to a belief system that denies the possibility of ESP. It is natural that many people who are not scientists follow the lead of well-known scientists who appear in the popular media, most of whom have embraced the doctrine of materialism.

- *The acceptance of psychic phenomena may require a more expanded consciousness.* It is likely that accepting the reality of psychic phenomena requires that a person have a wider, more universal or spiritual outlook on life. An analogy would be a blind person who one day miraculously gained sight. That person might begin to perceive a far richer, more beautiful reality with its promise of a new meaning and purpose of life, opening up a new realm of almost infinite possibilities. Similarly, someone who is mentally fixated on the material realm may find it difficult to appreciate or tune into a nonphysical ontology. As a result, no amount of evidence is going to change their worldview—a change in how reality is perceived or experienced may be a prerequisite.

PART III

SPIRITUAL IDEOLOGY

THE SPIRITUAL WORLDVIEW IS a top-down ontology that begins with Consciousness. It is Consciousness that is transformed into mind and then matter. From a blank-slate perspective, the top-down theory of existence has equal credibility to the bottom-up theory. However, from a scientific, logical, and metaphysical perspective the spiritual worldview is unquestionably more plausible than the alternative.

First, there is a need to explain how Consciousness, which is the subtlest thing in creation, is transformed into the material world. Secondly, we need to see what this worldview tells us about unit consciousness, mind, the purpose and meaning of life, karma, life after death, happiness, suffering, good and evil, etc. Thirdly, we want to see why this worldview explains all the anomalies of the materialistic worldview pointed out earlier as well as the weird discoveries of modern science relating to quantum mechanics, relativity theory, and evolution. Lastly, we want to answer the question of why the universe is so well tuned for our existence.

10
The Cycle of Creation

NUMEROUS SCIENTISTS ARE CONVINCED that the spiritual worldview is correct and have written extensively on the subject.[1] However, they have struggled to explain how Consciousness is transformed into mind and matter in a clear, simple, and rational way. This was not the case for Prabhat Ranjan Sarkar (aka Shrii Shrii Anandamurti). Sarkar was an Indian-born spiritual teacher, scientist, philosopher, neo-humanist, author, social revolutionary, poet, composer, linguist, social theorist, artist, and economist. He founded Ananda Marga (the Path of Bliss) in 1955, a spiritual and social-service organization that offers instruction in meditation and yoga. He wrote over two hundred books on various subjects, including history, spirituality, sociology, education, Tantra, yoga, medicine, ethics, psychology, humanities, linguistics, economics, ecology, farming, music, and literature, and has been recognized as one of the greatest philosophers and spiritual authorities of all time.

The spiritual ideology expounded by P. R. Sarkar provides a concise, rational explanation for how the cosmos evolved from Consciousness and how under the right conditions living organisms originated on Earth and evolved into self-conscious human beings. Finally, he explains how and why humans are subtly driven to reunite their individual or unit consciousness with the Cosmic Consciousness.

According to cosmologists, the universe began in a gigantic explosion originating from a dimensionless point (the Big Bang) prior to which there was no space or time. However, cosmologists have offered little beyond pure speculation on how the enormous mass-energy of the universe emerged from complete nothingness. Spiritual ideology, on the other hand, claims that the universe emerged not from nothingness, but from unqualified Cosmic Consciousness (Consciousness). However, without a force to qualify this Consciousness, nothing could be created from it. It would be like a lump of clay lacking the hands of a sculptor.

Hence, according to Sarkar and Eastern (Vedantic) philosophy, the Cosmic Entity, known as Brahma, has two complementary aspects: Consciousness (Sanskrit: *Purusha*) and a Qualifying or Creative Principle (*Prakriti*). Like other complementary constructs such as wave-particle, Consciousness and Qualifying Principle can be considered two sides of a coin—the coin representing Brahma. Without the action of the Qualifying Principle, the unqualified Consciousness could not undergo transformation and would remain unmanifest as pure awareness.

The Qualifying Principle has three basic binding modes that qualify differently. The subtlest mode is known as the sentient binding force (Sanskrit: *sattvaguna,* where *guna* means binding force). This creates the cosmic "I feeling." The second mode of the Qualifying Principle is the mutative or active principle (*rajoguna*). Acting on the "I feeling," it creates the cosmic "I do" feeling. The strongest binding force of *Prakriti* is the static principle (*tamoguna*). Acting on the "I do," it creates the "I have done" or "subjective I feeling." Before creation begins, the Cosmic Entity (Brahma) remains unqualified or in a state of potentiality or *Nirguna* (literally, without *guna*). The binding forces of the Qualifying Principle can be thought of as being disorganized or dormant and thus unable to act upon Consciousness.

In this model, creation is thought to begin when the three binding forces form an equilateral triangle and capture a portion of the unqualified Consciousness in the interior of the triangle. Once confined in this manner, Consciousness bursts forth from one of the vertices of the triangle, beginning the process of creation. This explosion of Consciousness from what is to become the emergent nucleus of creation is analogous to the beginning of the cosmos envisioned by astrophysicists. Thus begins the extroversive or centrifugal phase of the creation cycle, in which Consciousness undergoes a change from subtle to crude. *Saincara* is the Sanskrit name given to this particular phase of the creation cycle (*Brahma Chakra*).

The Consciousness (*Purusha*) that bursts from the emergent point of creation is not qualified but it does come within the scope of the Qualifying Principle. As such, the Cosmic Entity is said to be *Saguna* (literally, with *guna*). The unqualified Consciousness can be thought of as a wave with an infinite wavelength (straight line). This unqualified Consciousness is pure being or awareness without even the idea of its own existence.

When a portion of the unqualified Consciousness comes under the influence of the sentient binding force (*sattvaguna*), the cosmic "I exist" is created. This is analogous to creating a slight curvature in the straight-line wave of the unqualified Consciousness. This sense of "I exist" or "I am" (*mahattattva* or *mahat*) is nothing but the Cosmic "I." This is the first qualification or bondage of Consciousness, but it is very subtle and is imparted to only a microscopic portion of the unqualified Consciousness. When a portion of the "I exist" (*mahat*) comes under the influence of the mutative force of the Qualifying Principle (*Prakriti*), the cosmic "I do" (*ahamtattva* or *aham*) feeling is created. Naturally, the wavelength of the "I do" feeling is less than that of the "I exist" (*mahat*), since it is more qualified or bound. This Cosmic "doer I" is subjective in character and has no objective reality; like the *mahat*, it is more or less a theoretical construct and as such is outside the realm of science.

Dominance of the static force

As the static principle begins to dominate the creation cycle it first acts upon a portion of the cosmic "I do" feeling (*aham*) and creates the Cosmic "objective I" (*citta*), also known as "cosmic mind-stuff." This cosmic mind-stuff has objective reality but no physical reality, since matter is not yet formed. The cosmic mind-stuff is somewhat analogous to a mental construct. It could be compared to a scene we create in our minds. For example, close your eyes and visualize a person riding a horse. There is willfulness or doership associated with the sense of "I" as your mind takes on the colors and forms of the horse and rider. For us the scene has only subjective reality, much like a dream or hallucination. It has no objective reality since we lack the power to create a living horse and rider.

The transformation of Consciousness into "I exist," "I do," and "objective I" together constitute Cosmic Mind or the mind of God. This first phase of the creation cycle occurs even before the creation of space and time.

Under the gradually increasing static binding principle (*tamoguna*), a portion of the cosmic mind-stuff is transformed into space-time (*akasha*), also called ethereal factor. Recall that early scientists observed wave-like properties of light such as diffraction and thought that light waves reaching our planet from distant stars needed a medium through which to travel (just like all waves). They labeled this medium "luminiferous aether." The ethereal factor is not related to this outdated model; it simply refers to the four-dimensional continuum we call space-time.

Most nonscientists assume that space-time consists of a vacuum, which has no physical matter and no energy. However, quantum mechanics and relativity theory require that space-time contain a huge quantity of energy—much more than the total mass-energy of the universe. This is no small amount of energy since the total mass-energy, which includes dark matter, is very large. It is defined by Einstein's famous equation: $E = mc^2$; where energy (E) is equal to mass (m) times the speed of light in a vacuum (c) squared. Hence, space-time has tremendous energy locked within it and has particular significance for any understanding of the true nature of the cosmos. Interestingly, mystics have also claimed that the ethereal factor (*akasha*), which is the subtlest of the five fundamental factors, has the most energy. In addition, they say that it emits an OM sound, heard in the mind, which they attribute to the emergent sound of creation.

The transformation of Consciousness into "I feeling" (*mahat*), "doer I" (*aham*), and "subjective I" (*citta*) takes place outside the scope of time and before the start of the Big Bang recognized by cosmologists. The first formation of space-time marks the beginning of time or "time zero" for the Big Bang. In the first moments following time zero, space-time (*akasha*) begins to expand rapidly from the point of origin of the Big Bang (emergent nucleus of creation), resulting in a rapidly expanding physical universe. At the same time, the pressure of the static binding principle continues to increase and transforms a portion of the ethereal factor into enormous amounts of subatomic particles, which are constituents of what is called the aerial factor. Initially these particles composing the aerial factor are too hot to unite and form atoms, and this hot plasma emits radiation or luminous factor.

Evidence obtained from radio telescopes and the Planck spacecraft indicate that even today one can observe a relic of the Big Bang termed the cosmic microwave background radiation. Studies of this radiation show that it is not completely homogeneous. This may explain how matter eventually clumped together to form galaxies where the observable matter in the universe is concentrated. As the universe expanded and cooled, these subatomic particles, such as protons, electrons, and neutrons, coalesced and formed principally hydrogen gas. Hydrogen is the simplest element, having one proton and one electron. Small amounts of helium and lithium were also formed.

Under the relentless pressure of the static principle, the hydrogen atoms, which were not uniformly distributed in space, became attracted to one another by the pull of gravity. Once a huge mass of hydrogen forms, the gravitational force is so strong that it causes the hydrogen atoms to compress into incredibly dense plasma (atomic nuclei without electrons). Under these conditions of extreme heat and pressure, the nuclei begin to fuse, igniting a baby star. This process is termed thermonuclear fusion and is the same process that occurs in all stars, including our sun. It is also responsible for the enormous energy released by a hydrogen bomb.

The energy that is released when the hydrogen nuclei fuse to form a helium nucleus of two protons results from the loss of a small amount of mass, which is converted to energy. Even though the loss of mass is small, the energy is large. This process creates enormous amounts of luminous factor (*tejastattva*)—electromagnetic radiation (heat, light, X-rays, etc.).

Such electromagnetic radiation can travel no faster than the speed of light. If there were no speed limit for the physical components of creation, then in theory they could reach infinite speed. This would mean that for light, or even matter, they could be everywhere at the same time (nonlocal). Only mind has this capability. The speed limit for light is an insurmountable wall separating mind and matter that prevents matter from traveling to the past, which could create impossible paradoxes. The hierarchy Consciousness » mind » matter is maintained.

Solidified consciousness

A typical star consists principally of hot plasma, a form of luminous gas in which the atoms are ionized and do not have electrons directly associated with atomic nuclei. Besides helium, the second element, thermonuclear fusion creates small amounts of heavier elements, but none heavier than iron. For heavy elements such as copper, zinc, silver, gold, etc., to form, a massive star has to undergo a stellar explosion known as a supernova. Such events are rare but can emit as much energy in a few weeks as our sun will put out during its lifetime. In addition, supernovas throw off enormous amounts of gas and create the heavier elements, some of which are necessary for life. Hence, life as we know it would be impossible if we did not have within our bodies the dust from exploded stars. The shock wave from a supernova can also trigger the formation of new stars. Stellar material is constantly being recycled and telescopes reveal numerous nebulas in which new stars are observed forming from the dust of dead stars.[2]

As a star forms from gaseous hydrogen it will often develop what is termed a gaseous protoplanetary disk. After a substantial amount of time, hot gases in this disc cool with the accretion of dust and heavier material produced by the star or attracted to it by its massive gravitational field. Eventually these materials coalesce to form protoplanets. At first, a protoplanet may exist in a molten state, but over time it may cool and develop a solid crust. Thus the liquid and solid factors arise by the continued binding force of the static principle on the gaseous and luminous plasma that makes up a star. The solid factor is the crudest of the five fundamental factors. In other words, in a solid the pressure of the static force has reached its maximum.

In the solid state of matter, interatomic and intermolecular distances are at a minimum. However, this does not mean that solid matter is as solid as we perceive it with our sense organs. Almost the entire mass of an atom is concentrated in the tiny nucleus that holds the massive subatomic particles —protons and neutrons. The electrons exist in various shells or orbitals outside the nucleus and are maintained in position by their electrostatic attraction to the protons in the nucleus. However, the mass of an electron is only about one two-thousandth of a proton and the shells containing the electrons are many thousands of times greater in diameter than that of the nucleus. To put this in perspective, suppose the nucleus of a typical carbon atom was the size and weight of a lead BB. The electrons would be comparable to dust particles some seventy feet from the nucleus. Clearly, an atom is almost entirely empty space! The only reason the empty space composing your hand does not immediately pass through the empty space composing a table is that the outermost shells of the atoms in your hand and those in the table are filled with electrons. The electrostatic repulsion between the electrons in your hand and those of the table prevents your hand from passing right through atoms of the table. This creates the illusion of solidness, when in fact common matter is mostly just space.

Nonetheless, a solid object such as a rock or planet is ultimately composed of Consciousness—albeit in its crudest form. This is the nadir point of the creation cycle. Consciousness completes the outward or centrifugal phase of the cycle when solid material such as a rock or a protoplanet is formed. This is the inanimate phase of creation and represents the formation of Cosmic Mind and the macroscopic universe. The second, return phase of the cycle (*prati-saincara*) may commence at this point. It is the counter-movement or centripetal movement of the creation cycle. In this phase of creation, the unit or solid objects begin to evolve mind. The same layers of Cosmic Mind form but in reverse order, and they are contained within the physical boundaries of individual living organisms. The transformation of matter into living organisms takes place gradually due to constant struggle and the increasing reflection of Consciousness.

The return phase of creation

Life can only develop on a planet when there is a proper balance of the five fundamental factors: ethereal, aerial, luminous, liquid, and solid. On a planet like Mercury or Venus, life is not possible since there is an overabundance of luminous factor resulting in very high temperatures and no liquid water. On the other hand, frozen planets like Uranus and Neptune have insufficient luminous factor reaching their surface from the sun and are too cold for life to evolve. On gas giants like Jupiter and Saturn there is an overabundance of the aerial factor and essentially no solid or liquid surface for life to evolve. At one time Mars may have been hospitable for the development of living organisms since there is strong evidence that it once had abundant and flowing water and a robust atmosphere. Further exploration of Mars may prove that primitive life forms evolved there billions of years ago. However, today the lack of liquid water and a suitable atmosphere that would shield potential life forms from deadly solar radiation suggest that conditions are not currently conducive to life.

In our solar system, that leaves Earth where life did evolve. Earth is believed to have formed about 4.5 billion years ago. Initially a hot molten body, it was continually bombarded by meteorites and comets. The heavier elements such as iron and nickel settled into the core of the planet, where they are still found today in a molten state, kept warm by the radioactive decay of unstable heavy elements. The lighter substances gravitated to the surface of the planet and cooled, creating a hard crust.

Meteorites and comets containing liquid water and organic compounds including amino acids rained down on the surface of the infant Earth for hundreds of millions of years, creating oceans and continents by 4.3 billion years ago. A thick atmosphere formed consisting mainly of carbon dioxide, water vapor, sulfur compounds, methane, and nitrogen. The atmosphere helped Earth retain its liquid water and shielded potential life forms from deadly solar radiation. The conditions found on the surface of the infant Earth were conducive to the formation of life.

Life was bound to evolve on Earth because Consciousness is inherently creative and as such tries to express itself in the form of unit living beings whenever planetary conditions permit it. The first organisms that formed might have been proto life forms that became extinct long ago, but the oldest evidence of life on the earth is the fossilized remnants of cyanobacteria, dated at 3.5 billion years. These bacteria obtained energy via photosynthesis and therefore utilized energy from the sun. A byproduct of photosynthesis is oxygen, and it is believed that these early single-celled organisms were responsible for converting the early atmosphere, which was devoid of oxygen, to one containing oxygen. The oxidizing atmosphere was deadly to many of the other microbes that were present on earth at the time, but it was a vital component needed for the evolution of more complex organisms, including animal life.

In the absence of an organizing principle (Consciousness), the incredibly complex combination of atoms and molecules that make up even the simplest life forms would appear to be highly improbable. However, the Cosmic Entity relentlessly endeavors to express itself as living organisms in the return phase of the creation cycle. Thus, life will develop on a celestial body whenever the conditions allow for its expression. Since the universe contains countless stars, many of which have planetary systems, the universe must contain countless star systems that have extraterrestrial life. Within our Milky Way galaxy, the fraction of stars having planets is relatively small, and only a small fraction of planets observed to date have a solid crust and other conditions conducive for the development of life. There are probably only a small percentage of those planets where life developed and had favorable conditions long enough for the evolution of highly intelligent living beings. This, along with the vast distance between stars within a galaxy, makes it unlikely that we will ever make physical contact with intelligent beings from another world.

With the first spark of life on a planet, the return phase (*prati-saincara*) of creation begins. Only unicellular organisms populated the earth for a very long time (~3 billion years). However, even single-celled life forms express unit consciousness and possess a rudimentary unit mind. Naturally, they experience a constant struggle to survive. In this struggle for survival, the simple unicellular organisms found it advantageous to colonize with other cells and eventually multicellular organisms evolved. The time when multicellular organisms arose is difficult to determine with any precision but it may have been about one billion years ago. This was a giant evolutionary step; not surprisingly, it took a very long time before multicellular organisms began to dominate the biosphere.

Organisms evolve in the return phase of creation through the constant struggle for survival and the constant pressure of the Creative Principle (*Prakriti*). Struggle causes the unit mind-stuff to become more subtle. This leads to greater and greater expression of subtler mind-stuff, the "I do" and "I exist" factors (*aham* and *mahat*). Hence, organisms experience mental expansion, which is accompanied by greater physical complexity. Life forms gradually evolve into higher and higher species such as fungi, plants, invertebrates, vertebrates, and finally mammals.

The return path toward the starting point of the creation cycle (Consciousness) is not without starts and stops as many life forms come and go when they fail to adapt to changing conditions. However, the movement from crude, less developed life forms to mentally subtler and physically more complex life forms is inevitable because like a cosmic clock the hands of evolution always move forward under the prevailing pull of the Cosmic Entity (attraction for the Great).

The development of higher mental functions

Once the unit "I do" feeling (*aham*) in an organism becomes strong enough, intellect begins to develop. In Sanskrit, the word for intellect is *buddhi*, which is related to Buddha, meaning the "Enlightened One." In less evolved animals, the cosmic mind-stuff is reflected in the unit's rudimentary mind. The animal is principally guided by instincts. However, because its instinctive mind is connected with cosmic mind-stuff (a nonlocal collective mind), those instincts do not arise in the physical brain of the organism and do not need to be passed on by heredity.

As more and more "I do" feeling (*aham*) develops in a creature, it begins to learn from its experiences and modify its behavior accordingly. For example, a dog learns a series of tricks through training. This learned behavior involves some degree of intellect, since this behavior is not instinctual for the dog. Most animals show some degree of intellect, since they learn from experience and modify their behavior based on those experiences.

Higher animals such as apes, dolphins, whales, seals, etc. have a fairly well-developed intellect or intelligence. They are able to solve problems using abstract reasoning. That is, they exhibit behavior that is not trial and error but seems to involve the ability to apply previous knowledge to a new situation. Hence, there is a gradual transformation of higher mental functions in animals and not a quantum leap from apes to early hominoids. This "I do" capacity of mind or *aham* grows by the constant psychic struggle with stressful situations that force the organism to adapt and grow mentally. For example, a young deer wanders down to the river for a drink, but a crocodile is waiting submerged, hidden from sight. As the deer leans over for a drink of water the crocodile leaps toward it, jaws open. The deer escapes almost certain death when it jumps back just before the jaws of the crocodile are about to close on its leg. From this frightening experience, the deer learns to be more cautious when approaching the river for a drink. As the mind develops, the physical structure of the organism becomes more complex. The struggle for survival creates conditions that lead to adaptive changes that are passed on to subsequent generations, just as Darwin hypothesized in his theory of evolution.

Animals appear to have some degree of cognitive interconnectedness. A new learned behavior is passed along to other animals of the same species via their connection with Cosmic Mind. Because instinctual behavior is guided by Cosmic Mind (principally cosmic mind-stuff), learned behavior of individual animals can theoretically be passed on to others, helping to advance the evolutionary development of the species.

Evolution of man

The relentless march of evolution is slow. It has taken nearly a billion years for intelligent life to evolve from the first multicellular organisms. Modern humans evolved from their hominid ancestors about 250,000 years ago. But how do human beings differ from other animals? As we have seen, the unit mind-stuff is increasingly transformed into "I do" (*aham*) and "I am" (*mahat*) as creatures develop mentally on the path of *prati-saincara*. The individual or unit mind grows in proportion to the reflection of the Cosmic Mind on the unit mind. As the mind expands, the physical structure becomes more complex, with more and subtler glands, in order to adjust to the higher psychic sentiments and demands. The ego, or sense of doership, develops, followed by an increased sense of self-awareness. Once the sense of "I exist" or *mahat* becomes predominant, the organism is fully self-aware and self-determinant. We call such creatures human beings.

Since humans have a developed sense of self or ego, they possess free will and can move their mind according to their desire. Plants and animals, which lack the developed sense of "I exist," cannot act independently. They act according to instincts. Being so guided, lower forms of life do not possess the ability to go against the natural flow of the creation cycle. Man, however, has the ability to focus his mind in any direction he chooses, and mind always takes on the qualities of the object of its attention. This is a double-edged sword since man can choose to focus on the subtlety of Consciousness and move forward on the path of evolution at an accelerated pace, or he can choose to direct his mind toward the crude and move backward toward the unconsciousness of animal existence.

As organisms developed physically, they evolved sense organs in order to better interact and adapt to their physical surroundings. With the exception of the ethereal factor, which can only be sensed mentally, organs developed that allowed organisms to sense the other four fundamental factors (aerial, luminous, liquid, and solid). The five principal senses are smell, taste, sight, hearing, and touch. The seats for the five sensory organs are actually in the brain. For example, we see our outside environment on the visual cortex inside our brain. The gateway for sight might be the eyes, but no light enters the brain, only nerve signals from the optic nerve. This is true for the other four senses as well. The sense of hearing and touch detect the aerial factor. The luminous factor is detected with the eyes and by the sense of touch (heat). The liquid and solid factors can be sensed by the other sense organs and often by smell and taste.

The mind is able to translate the nerve impulses reaching the brain from the gateway organs into externally projected reality with the help of mind-stuff, which in a sense takes on the form of the sensory vibrations reaching the brain. Similarly, when one experiences thoughts and daydreams, the unit mind-stuff is transformed in the process. Hence, the mind always takes on the properties of the object of its attention, whether internal or external. When the mind is involved with sensing and reacting to the external environment, or thinking about material objects in the external world then it is not able to expand and become subtler.

Human beings are dominated by unit "I feeling" and are naturally attracted to the Cosmic Entity. The difference between the unit consciousness and the Cosmic Entity is that the unit beings are multi-purposive and unilateral while the Cosmic Entity is multilateral and unipurposive. Units are multipurposive because they desire many things, but they are unilateral because they can do only one thing at a time. The Cosmic Entity is multilateral because he guides and witnesses everything in creation, but he is unipurposive because he has only one desire—that all his created beings return and merge with him. Hence, as hard as human beings may try to go against the subtle pull of Consciousness, they are relentlessly drawn toward the Cosmic Entity. In the final analysis, it is impossible to go against the natural flow of the cosmic cycle. According to the model, all living organisms will eventually attain human status and all such units will eventually merge with the Cosmic Entity.

The final journey

The final step in the cosmic cycle is the merger of the unit consciousness with the Cosmic Consciousness. Such merger is a return to the starting point of creation. Merger occurs as the individual mind recognizes more and more that it is an inseparable part of the Cosmic Mind. In the process, the individual mind is lost—like a grain of salt dissolving into the ocean.

The model for the creation cycle depends on the concept of reincarnation, since the ultimate dissolution of the unit into the Cosmic Entity can take more than one lifetime. One's unit consciousness or atman can be considered to have begun billions of years ago in the form of a single-celled organism and then evolved gradually in the return phase of the cycle before attaining human status. Then it may take many lifetimes before spiritual union is obtained. The story of evolution proposed by Darwin would agree that physically and mentally we are the product of billions of years of evolution, but materialist ontology does not recognize that something nonphysical (i.e. the unit mind and consciousness) continues uninterrupted as the physical state of an organism changes (dies and is reborn). We will discuss the subject of reincarnation in the next chapter.

For human beings, the path of knowledge is called *vidya* or *ananda marga* (path of divine bliss) in Sanskrit. *Vidya* also means force or tendency for good and its opposite, *avidya*, means evil force or tendency. When an individual finally realizes that their true purpose in life is to try to merge with the Cosmic Entity, then they may decide to embark on a path leading to that goal. Spiritual practices including meditation are tools that help one move forward on this path. Spiritual practices help transform the ego into "I am" feeling (*mahat*). This can be both difficult and painful, and at first one's growth may be measured more by the difficulties experienced and overcome than by the pleasure one feels. Embarking on the path of *vidya*, the mind begins expanding more rapidly and becomes increasingly subtle. There is an accelerated movement of the unit mind toward the ultimate goal. It is like a small fish living in the stagnant waters of a pond that leaps a dam into a tiny brook. At first, the flow is a trickle but soon it increases exponentially to that of a great river, and eventually that river reaches the ocean. Hence, the path is ultimately blissful as the individual feels that they are finally moving toward their goal, and the experience is one of greater and greater happiness as they draw closer to that goal.

Yoga is one of the paths that lead to the goal of unification, but there are many others. Yoga means "union" and should not be confused with yoga postures or asanas, which are only a small part of the discipline. Paths vary according to the person, time, and place. Just as there can be many paths leading to a mountain peak, so there are myriad roads or practices leading to the realization of the Cosmic Entity. However, the path taken will not be easy or without difficulties.

Many human activities beside meditation can produce mental expansion and hence fall under the category of spiritual practice. Any activity in which the ego is set aside and the mind is absorbed in the creative flow of invention, problem solving, artistic creation, or other such mental activities can produce mental expansion similar to meditation. Such activities can generate pleasure, peace, intuitional knowledge, and diminish the ego. Probably the most powerful types of spiritual practice are those where there is a willful turning of the mind inward toward the Cosmic "I am" or Cosmic Entity. This type of meditation practice greatly accelerates the natural movement of the unit consciousness toward its goal of merging with the Cosmic Entity.

The opposite movement, termed *avidya* in Sanskrit, is also possible for humans. If the mind is strongly attached to physical objects and desires like money, pleasure, name, fame, etc., then it takes on the qualities of these objects and desires and becomes increasingly crude. If the mind degrades to a point where it loses the human traits of compassion, conscience, fairness, etc., it may no longer be able to maintain parallelism with a human body. Theoretically, such a soul could be born into an animal body. This movement from subtle to crude goes counter to the natural flow of the creation cycle and can only be suffered by human beings because they are free to choose how they focus their minds. This is why free will is a double-edged sword. It confers in humans the ability to do spiritual practices and greatly accelerate the pace of evolution, but at the same time it makes it possible for a human being to regress rapidly and possibly return to animal existence. If this occurs then it is a great tragedy, since it has taken millions of years for human beings to evolve from their distant ancestors. However, the wheel of the creation cycle turns inexorably and they will eventually regain their human status.

Obviously, there is purposefulness at work in the creation cycle. In some sense, the Cosmic Entity has allowed himself to come under the influence of the Qualifying Principle and take on the myriad forms of the beings that are inescapably drawn back to the Source that created them. If we use the analogy of a clock, then human beings are at the 11:59 point in this cycle. Whether one consciously realizes it or not, they are constantly being pulled toward finishing the cycle and merging their little "I am" into the Cosmic "I" and thus into the unqualified Consciousness.

This is the true purpose and meaning of existence; to deny this purpose would be to throw oneself backward toward the darkness of unconsciousness. However, the question remains, why did the Cosmic Entity allow a part of his being to come under the bondage of the Qualifying Principle in the first place? There is no way to answer this question. Some call it his *liila* or play. They say he was bored because there was no play to witness. Was he really? We will never know the answer to this question without becoming one with him.

THE CYCLE OF CREATION

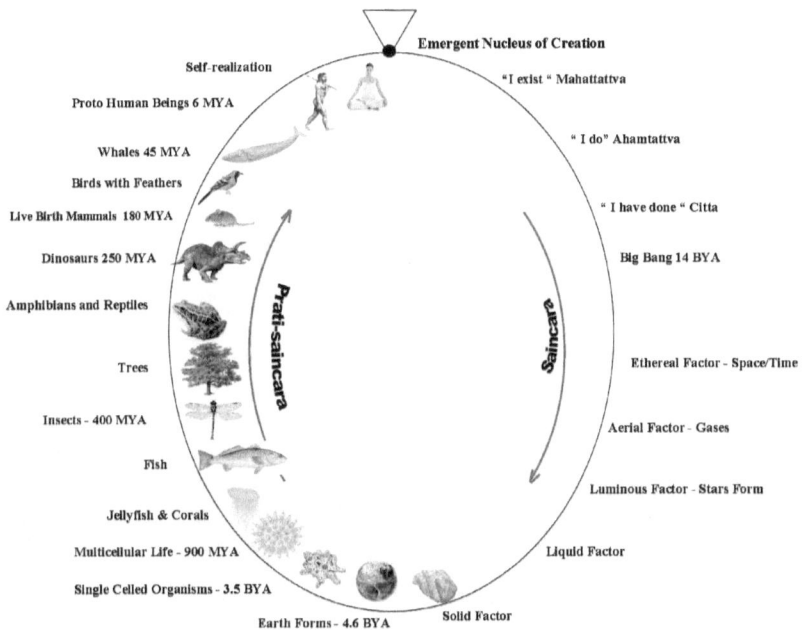

The creation cycle (*Brahma Chakra*).

11
Mind, Body, and Death

THE SPIRITUAL MODEL FOR creation explains how Consciousness is transformed into Cosmic Mind, then into space, time, energy, and matter, and finally into living organisms. It describes evolution as teleological in that it moves toward an ultimate goal of reuniting the unit with the Cosmic. It says that there is meaningfulness and purpose for human existence. It says that mind is nonlocal and that all living organisms have a mental link to the Cosmic Mind and that this Cosmic Mind is transcendent, e.g. beyond space and time, and is witness to the entire creation from an outside-in point of view. In other words, the entire creation is an internal thought projection of the Cosmic Entity.

Consciousness is the ground substance for creation and is therefore neither created nor destroyed—it is eternal, unbound by time. Since the unit consciousness represents a minuscule portion of the Cosmic Consciousness, it too is eternal, beyond the bondages of space and time, and is only "extinguished" when it becomes one with Cosmic Consciousness. However, extinguished may be the wrong word for this process, since upon its merger into the ocean of pure Consciousness, the unit becomes much, much more than it was before. An analogy might be the reflection of the sun in a mirror—the mirror representing the unit's mind. Upon merger, the mirror is dissolved and the experience is that of the sun itself—in contrast to a minuscule portion of its radiation. In reality, we are no more separate from the Source than rays of sunlight are separate from the sun.

The model describes reality as one or singular. Its nature is wholeness rather than consisting of individual parts. Everything is intimately connected and interdependent. Differentiation, individuality, separateness are ultimately an illusion created by an inability to perceive the big picture. For humans, awareness is focused on mind and body, and from a young age people learn to differentiate between self and nonself. This self-image, or ego, is needed in order to function in the world, but it also creates a barrier to understanding one's true nature. As a result, people experience a relative reality; rarely do they dive deep within to experience the oneness, the Ultimate Reality.

The spiritual model for creation asserts that human beings are very close to the end of the cycle, but that they need to make a conscious effort to complete the journey. If they simply ride the waves on the surface of their being, never diving deep within to discover their connection with the limitless, infinite Cosmic Entity, then they progress very slowly toward the goal of merger.

Actually, the "I feeling" is nonlocal and universal. Imagine a radio station broadcasting music from the top of a hill. The radio waves propagate in all directions and are received by the radio in your car. Many other cars also tune into the station, and although the cars all move around each picks up the same music. Similarly, there is actually only one "I feeling" being broadcast in the cosmos, the Cosmic *Mahat*. The unit minds act like radio receivers. They pick up the vibration of the Cosmic "I" and translate it into "I am Adam" or "I am Patricia," etc. If this one "I" was not at work processing all the sensory data every second in every brain, then it is difficult to see how people could experience the same reality. Each brain is different, yet there is a coherent experience among all people that can only be explained by the fact that a single Consciousness is at work. It is not enough to say that we see the same reality because reality is "out there." The fact is, our only experience of reality is in our brain; what is outside is secondary to our experience.

If through spiritual practice or sudden realization one experiences their own, unqualified "I am" feeling, then they become aware of how it transcends their little sense of self. They realize that this same "I feeling" is experienced by every living creature. This experience is sometimes called self-realization or *savikalpa* samadhi—the realization "I am God."

Layers of mind

Spiritual ideology teaches that the body and conscious mind are the crudest layers of mind and are dominated by the static principle. Here the physical body is termed the *annamaya kosha* in Sanskrit. *Annamaya* refers to the body created by food and *kosha* means "layer of mind." The needs of the physical body come first, and it is not until those needs have been satisfied that individuals are able to devote the time and energy necessary to enter the path of enlightenment.

The conscious mind (*kamamaya kosha*) is mostly involved with the physical body and controls the motor organs. It also receives input from the sense organs, which are actually based in the brain. This layer of mind is active when one is awake. It is home to desires, fears, and pleasures. It is the crude mind because it is involved with sensing and acting in the external physical world. The conscious mind governs the sense and motor organs, and it deals with bodily needs such as eating, drinking, sleeping, and procreating. It experiences fear and is involved with self-preservation. Man shares with animals the basic needs and desires controlled by the conscious mind. The only difference is that animals are primarily driven by instincts and are incapable of performing higher mental activities, such as contemplation of God.

The third layer of mind is the subconscious mind (*manomaya kosha*). The mutative principle dominates this layer. It is active all the time except in deep sleep, and one is most aware of it when dreaming. The subconscious mind does a lot of its work nonverbally—that is, using images instead of words. Being subtler than the conscious mind, it is more expansive and is involved with memory, comprehension, imagination, and the higher mental functions such as philosophical thought, scientific reasoning, and information management. It sorts memories and creates mental pictures and one is rarely aware of its activities. It is an untiring slave to the conscious mind, working constantly behind the scene to solve problems and prepare one for success. Lying between the conscious and unconscious layers of mind, intuitive ideas and solutions to problems often pass through it and emerge into waking consciousness. Sometimes a person can go to bed with a problem and wake up having found the solution. Many intellectual and scientific discoveries have occurred when the conscious mind and the sense organs were in a quiet state, allowing the clarity that can provide insights into solving problems.

Most memories lie in this layer of mind. When activated, memories are expressed in the conscious layer. Memory is a complex process and may consist of exquisite details including mental pictures, sounds, tastes, and smells—sometimes from a third person point of view. The mind rather than the brain is the seat of memory. The unit mind-stuff records experiences and knowledge taking the shape, so to speak, of the event or information. Replay of a memory is dependent on a healthy brain to translate these mental vibrations into the electrochemical processes of the nerve fibers. Hence, if the brain is damaged, the mind may be unable to express memories in the conscious layer of mind. However, since memory is not simply a brain function, memory of events can sometimes be recalled with exquisite detail following hypnosis or electrical stimulation of the brain. Since memories are stored in the mind and not the brain, memories remain in the unit mind-stuff following death. Thus, it follows that memories and abilities that were developed in a previous life can be experienced and utilized in a later life.

The fourth layer of unit mind is the unconscious mind—sometimes called the superconscious mind. This layer actually consists of three separate layers. The yogic terms for these layers are *atimanasa*, *vijinanamaya*, and *hiranmaya*. The unconscious mind is noncerebral. It is an all-knowing collective mind shared by all humans. It is not bound by time, place, or person. It is the transcendent, omniscient, timeless collective mind, shared by all unit entities. When it is experienced, it is nothing but the reflection of Cosmic Mind on the individual's mind. The unconscious mind is responsible for all higher sentiments and knowledge, including magnanimity, humility, serenity, gentleness, mercy, intuition, psychic knowledge, discrimination, non-attachment, imperturbability, creative insight, and spiritual ecstasy.

Carl Jung called this level of mind the "collective unconscious." His evidence for the existence of this universal mind linking all humankind was based on the observation that among cultures with no historical influence upon one another, the basic symbols and myths were universal. Jung called the predisposed patterns for particular myths and symbols "archetypes." He argued that they exist in the collective unconscious of humankind and are similar to animal instincts in that they influence the basic structure and organization of the human psyche. Jung identified several archetypes, such as the hero, the mother, and the devil.

He also offered numerous examples of myths and symbols that arose in cultures with no direct link to one another. One example of a pervasive symbol is the cross and another is the twisted cross or swastika. The word "swastika" is derived from Sanskrit: *su* (good) + *asti* (being) + *ka* (object). Hence, it symbolizes well-being, good luck, or spiritual success. Besides being an ancient symbol of Hinduism, Buddhism, and Jainism, it has been found in Neolithic Europe, Asia, Africa, the Middle East, and China. It was also used by Celts, Greeks, Romans, Germanic and Slavic tribes, and by Native Americans. Archaeological evidence suggests that it served as a good luck charm or religious symbol in those various cultures.[1]

Joseph Campbell also provides many other examples of universal symbols and myths in his book *The Hero with a Thousand Faces*.[2] The existence of a collective or universal mind provides a simple explanation for why these archetypal patterns exist. It also provides a mechanism for ESP and a vehicle for the continuation of memories and capabilities from one lifetime to the next.

Beyond the unconscious mind lies the self or atman, which is the witnessing entity of individual being. Sometimes this immortal self is called the soul. Experience of the subtle unconscious mind gives expanded awareness, knowledge, and energy. It is rare for an individual to penetrate deeper than their subconscious mind, but if they do, it is a profound and moving experience. For a brief moment, they may experience the uninhibited flow of ecstasy, supermundane knowledge, and unconditional love present in the unconscious mind. After a while, the unrelenting attraction for the physical world inevitably draws them back to "normal consciousness." However, this firsthand mystical experience will forever change the way they view reality.

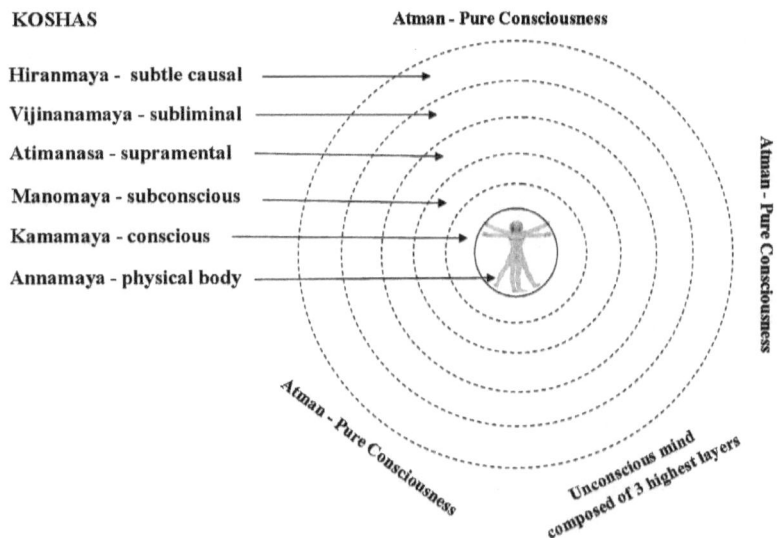

The six layers of the mind. The atman is the unit consciousness and is witness to all layers of the mind.

Psychic body

True power lies in the subtle, not the crude. As we have seen, the energy locked up in the mass of the universe pales in comparison to the energy of the void (space-time). Scientists call this "dark energy" and have calculated that it accounts for roughly three-fourths of the total mass-energy of the universe. Similarly, a person may be able to lift a barbell weighing two hundred kilograms, but if a tiny nerve is cut, they may be unable to lift a cup of water. Muscles depend on nerves, which in turn depend on the brain. And what is the brain dependent on? The mind or psyche.

Everything in the universe originates from Consciousness; it is the source of all power, energy, and matter. In the hierarchy of Consciousness followed by mind and then matter, there is diminished power and energy associated with the cruder factor. This is also true of the fundamental factors and explains why there is more energy in the vacuum of space-time than in all the matter of the known universe.

Every living organism, even the most primitive, displays traces of mind and consciousness. For example, a paramecium (a single-celled protozoan) can be observed in a microscope to swim about swiftly searching for food. If it bumps into an object, it recoils and darts off in another direction. Similarly, euglenas (unicellular protist) have an eyespot, a primitive organelle that is sensitive to light, and they are able to adjust their position in order to produce more food via photosynthesis. More evolved multicellular organisms show greater and more complex mental capabilities.

For a more evolved organism such as a cat with trillions of individual cells, it is difficult to explain how this very complicated physical system with brain, neurons, sense organs, autonomic systems, and numerous glands can function in an organized and coordinated way. Similarly, it is difficult to understand how all the different structures and systems develop in an embryo beginning with a single fertilized ovum. All the cells have the exact same DNA but somehow they know exactly when and how to differentiate into cells that are to become bone, nerve, muscle, etc. There is exquisite control over how cells develop in the embryo. Absent an organizing "field" or "force" it is hard to imagine how the incredibly fine-tuned development and other complex activities required for life could occur.

According to spiritual ideology, the unifying force that allows complex organisms to function and survive is their psychic body. This psychic body is their connection to Cosmic Mind, and it is through this connection that their physical body with its innumerable individual cells is able to form a viable organism. By being a bridge between Cosmic Mind and the physical body, the psychic body serves as the vehicle by which instincts and various autonomic functions are expressed.

In the West, the concept of a psychic or "vital" body was largely dispensed with by the biological sciences as they moved toward reductive materialism as an explanation for all life processes. However, things have begun to swing back with the publication in 1981 of Rupert Sheldrake's ground-breaking book *A New Science of Life*.[3] He argued persuasively that there exists a nonlocal, nonphysical "morphogenic" field that is responsible for the form and organization of the trillions of individual cells that constitute a living organism, without which it could not function or develop from embryo to adult.

The psychic body of a living organism is closely related to its physical body. For every physical organ, there is a corresponding psychic one. In vertebrates, the most important controlling nerve centers or plexuses of the physical body lie along the spinal cord and in the brain. Psychic centers are closely related to these physical nerve plexuses. They are called chakras in Sanskrit, which literally means "circles." Seven chakras have been identified in the human body by yogis. These are: (1) the *muladhara* chakra at the base of the spine, above the perineum; (2) the *svadhisthana* chakra, controlling the genital organs; (3) the *manipura* chakra at the navel or solar plexus; (4) the *anahata* chakra, associated with the heart; (5) the *vishuddha* chakra at the throat; (6) the *ajina* chakra, located between the eyebrows and controlling the mind; and (7) the *sahasrara* chakra at the crown of the head, which controls all the other psychic centers.

It is believed that the first five chakras are the controlling points of the five fundamental factors within the human body. The *muladhara* controls the solid factor, *svadhisthana* the liquid, and so forth. The unit mind is a minuscule clone of the Cosmic Mind and is controlled by the *ajina* chakra. The highest chakra, the *sahasrara*, is called the thousand-petaled lotus because it controls the other chakras and the one thousand propensities of the psychophysical body. This chakra is also the seat of pure unit consciousness and one's connection with the Cosmic Entity.

Not only are the chakras each associated with corresponding physical nerve centers or plexuses, but also with major glands of the endocrine system. They are also associated with other glands like the liver, spleen, pancreas, etc. They control the flow of psychic energy in the body. The nonphysical, psychic nerves of the body are termed *nadis* in Sanskrit and the psychic or vital energy they transmit is termed *prana* (Chinese folk medicine calls this chi).

Each of the chakras has a color, certain sounds, and a number of flower-like petals associated with it. These fundamental sounds number fifty in all, corresponding to the fifty vowels and consonants of the Sanskrit language. Each petal is believed to control one of the fifty major human propensities, such as anger, hate, desire, love, etc. The five *koshas* or layers of the mind control the five lower chakras (*muladhara, svadhisthana, manipura, anahata, and vishuddha*). The greater the control one achieves over one's psychic body and chakras through spiritual practice, the more control one obtains over their organs. Thus yogis who spend many hours performing asanas (yoga poses) and meditation may be able to gain extraordinary control over their bodies and minds as they become established in the higher levels of mind.

The base or *muladhara* chakra controls the crudest factor—solid. Solid represents the lowest point in the creation cycle. It is the point where the Qualifying Principle is strongest and where the potential for spiritual growth and expression is greatest. Thus the *muladhara* chakra is thought to be the home of the kundalini, the dormant spiritual potentiality. It is believed that spiritual practices can awaken this serpentine-like energy, and as it passes upward through the principal psychic nerves (*nadis*) in the spinal cord, it pierces the various chakras and progressively leads to greater and greater divine experiences. When the kundalini reaches the uppermost chakra, the *sahasrara*, the unit being unites with the Cosmic Entity.

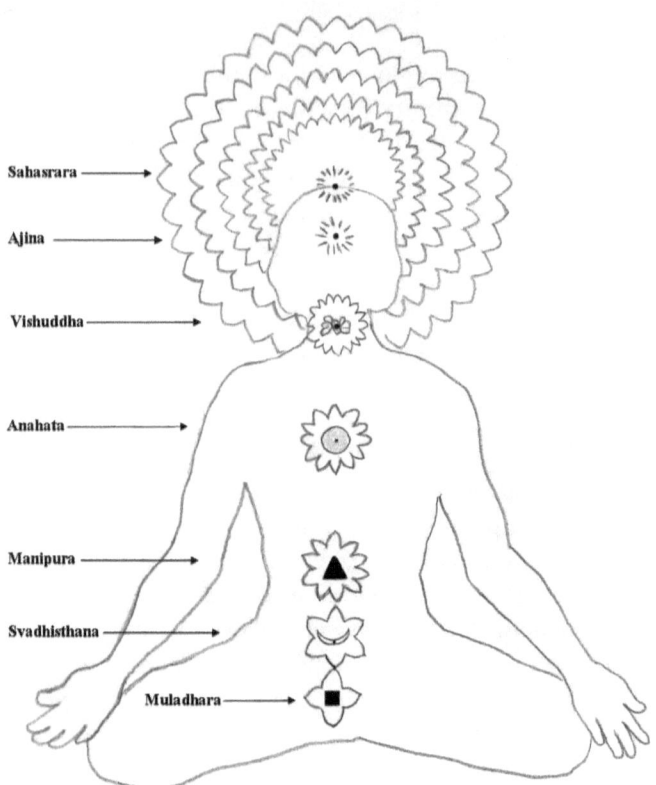

The seven principal chakras of the human body.

It is apparent from the above discussion that the psychic body has tremendous spiritual significance. Only human beings possess the extremely complex psychophysical structure that enables them to be fully self-aware and perform the spiritual practices necessary for moving on the *vidya* path to perfection. Animals have psychic bodies but their physical brains and corresponding psychic structures are more primitive and do not support a high degree of self-awareness. Therefore, animals are unable to perform spiritual practices.

It is with the help of the crude physical nerves and corresponding psychic body that the mind connects and activates the brain, thereby activating the body. Without the nerve fibers, it is impossible for the mind either to connect to the physical body or to function physically. Brain is necessary for the normal expression of mind, but mind is nonlocal and not dependent on a nervous system for its existence. During death both the psychic body and the physical body dissolve. The bodiless mind of a dead person, in the absence of the psychic body and physical nerve cells, is unable to express thoughts or emotions, or to affect anything physically. Thus the bodiless mind cannot contemplate the Cosmic Entity, express its hopes and desires, or become involved with any living entity or crude object. It must become associated with a new physical body in order to progress on the path to perfection.

Reincarnation

The most fundamental tenet of spiritual ideology is that Consciousness reigns supreme. It has no beginning or end. It is neither created nor destroyed. It simply is, and will be forever. All living beings possess consciousness, and this eternal aspect of their existence confers immortality upon every creature in the universe. The physical body may fall sick and die, but the unit consciousness that was associated with that physical body can never be extinguished unless and until it merges and becomes one with Cosmic Consciousness.

Thus, life after death is not a "pie in the sky" concept promoted by the religions of the world but is both a necessary and sufficient requirement of the spiritual worldview. Reincarnation is a corollary to the spiritual worldview, and it is supported by considerable scientific evidence, for example, by people who remember their past lives.

What actually happens in death? The simple explanation is that the mind loses parallelism with the physical body. When an organism is alive, there is a direct connection between its unit mind and its body. Consciousness is able to operate in the body. In higher organisms, the unit mind functions through the nervous system and brain. If the brain is damaged, then expression of the unit mind may be impaired or cut off. Physical death occurs when the unit mind can no longer maintain connection with the physical body. This normally occurs during brain death. At this point the unit mind is called a bodiless mind.

The bodiless mind retains all the memories and reactions to past activities and experiences. These mental impressions or reactive momenta (*samskaras*) are stored in the unit's mind-stuff and are noncerebral because mind-stuff is nonphysical and not dependent on the physical brain. The bodiless mind cannot feel pleasure or pain. Having no sense or motor organs, it is incapable of sensing or doing anything. It does not actively witness anything and exists in a state similar to deep sleep.

While Newton's third law describes the law of action and reaction for physical bodies—for every action, there is always an equal and opposite reaction—a similar law applies to the mental realm. It is called the Law of Karma. However, there is one major difference between action and reaction in the mental sphere—the reaction may be stored in the mind and experienced later.

Every physical and mental action involves the unit mind-stuff. When the mind-stuff is vibrated, it leaves an impression or reactive momentum (*samskara*). Such a stored reactive momentum is the potential for future action. When the stored reactive momentum is expressed, it is said to be "burned" or exhausted. Karma is nothing but the totality of all these reactive momenta. The burning of a reactive momentum returns the mind to a "cleaner" state. This process of making and burning a reaction is analogous to making a dent in a hollow rubber ball with your finger. The depression may last for a while, but if the ball is warmed with the hand, the dent can pop out, creating a perfectly symmetrical ball once again. In this example, the dent represents the reactive momentum. When the dent pops out, it is burned.

Reactive momenta result from mental and physical actions performed by the individual. They are stored as potential mental energy in the individual's mind, and their expression or burning is accompanied by the release of kinetic mental or physical energy equivalent to the mental energy or impression that created it. Although the amount of mental energy may be the same, the type or quality expressed will normally be different. For example, a person does a kind deed for a stranger on a road, such as helping them change a flat tire. The reaction to such kindness will be stored in their mind and perhaps come back to them later in the form of help someone else provides them. However, it will probably not be help changing a flat tire. In a subtle, often unconscious way, the reaction teaches the individual that it is good to help and be nice to other people.

Similarly, a bad deed will ultimately bounce back, creating pain, unhappiness, or suffering. That mental disturbance reminds the individual either consciously or unconsciously that the action that created the reactive momentum was bad, and they will be less apt to repeat that action. This is why the Law of Karma is often described as "you reap what you sow";

or "what goes around comes around." It is the basic ethical rule that most people live by. It is the basis for the Golden Rule, to do unto others as you would have them do unto you. The law of action and reaction is a reward/punishment system by which people learn to be better individuals, develop mentally, and ultimately gain wisdom.

Humans are free to direct their mind according to their will and thus may choose to act according to their own whims. Therefore they are free to create both good and bad reactive momenta. However, they are not free as to when and how they are burned. The greater the mental vibration or intensity of an action, the more powerful will be the associated reaction and the stronger will be its effect when the reactive momentum is experienced. Strong desires create a strong potential for action. For example, a boy witnesses firefighters extricating a person from a badly damaged car and develops a strong desire to become a firefighter himself. After many years of training, he may satisfy this urge to help people.

Actions performed unconsciously or with little or no thought do not create reactive momenta. There is no reaction if there is no "I do" feeling (*aham* or ego) involved. Therefore the action of others, actions performed unconsciously, and acts of God, such as floods or windstorms, do not create reactions in one's mind directly. However, such experiences may create other reactions. In Chapter 14, we will see how actions can be performed without creating new reactive momenta.

In order for an individual to express unburned reactive momenta following the physical death of the body, the Cosmic Mind will have to find a new body for the unit mind. This could take place almost immediately or it could take many years. Once the Cosmic Mind finds a suitable home for the bodiless mind, it will become associated with a fertilized ovum or zygote and eventually mature into a baby and be reborn. Thus the minimum time between death and rebirth for humans is about nine months. The same process applies to animals, except that they are guided by instinct and not willful intent. Animals are incapable of performing actions that could be judged as good or bad. They do not suffer from the consequences of their actions and upon death simply take on a slightly more advanced psychic and physical structure in accord with the inexorable movement of the creation cycle.

Since noncerebral memories can include memories of past events, including those that took place in previous bodies, it is sometimes possible for persons to recall experiences from a previous life. As we have seen, children are more likely than adults to recall memories of past lives. For children this can be a cause of anxiety, and in the West their parents will usually try to assure them that these memories have no basis in reality. It is fortuitous that these memories of living in another body fade by the time they reach ten years of age.

For adults, remembrances of past lives may come out during dreams, deep meditation, or during hypnosis. As mentioned earlier, a common clinical treatment for persons suffering from phobias or neurotic fear is hypnotic regression to a time when they first experienced an incident associated with the intense fear. Quite often, the traumatic experience they describe under hypnosis occurred in another body. Reliving this experience under hypnosis can cause a catharsis and eliminate the phobia.

Contrary to popular belief about reincarnation, the vast majority of cases involve people recalling normal, unexciting lives. Except for medical treatment, the remembering of past lives is not recommended since it can create anxiety and uncertainty, and can divert one's attention from the job at hand, which is to know one's self in the here and now.

Psychic abilities

All psychic abilities, including telepathy, clairvoyance, precognition, and psychokinesis, depend on the fact that the subtler layers of mind are nonlocal. By tapping into the unconscious layers of mind, one can know and experience things beyond the sense organs and thus beyond brain activity. Psychokinesis might be the most mysterious psychic ability, because it involves moving or influencing physical objects or equipment with the mind. As we have seen, the mystery of how the nonphysical mind can influence a physical object is no different from the way the mind moves the hand. As pointed out before, this interaction requires an act of observation (consciousness). A quantum particle or system can simultaneously exist and not exist; exist in several different states at the same time; or be spread throughout the entire universe. This is required by the uncertainty principle and by the wave function describing the particle or system. However, when the particle is measured or observed by any means, its state is no longer uncertain, and this act of measurement or observation instantly forces it into just one state. Experimental studies have proven repeatedly that consciousness or mind affects the fate of quantum particles by fixing them in a particular state. When the quantum wave function is collapsed by mind or consciousness, there is a fixing of matter in momentum, energy, position, or space-time. This observation led physicist Bernard d'Espagnat to write the following in an article in Scientific American:[4]

> The doctrine that the world is made up of objects whose existence is independent of human consciousness turns out to be in conflict with quantum mechanics and with facts established by experiment.

The mechanism by which mind or consciousness influences matter would account for not only how mind affects brain but also how mental intent can affect physical objects and sensitive electronic devices. According to the spiritual worldview, this interaction is made possible by the fact that matter is an epiphenomenon of mind, i.e. a product of mind, and not vice versa.

It is a common belief among those schooled in spiritual philosophy that spiritual practices can produce psychic or paranormal powers. These may include various abilities such as levitation, psychic travel, etc. These powers or siddhis are not the goal of spiritual practice and may actually distract one from the goal, which is merger with the Cosmic Entity. Such powers would depend on one's ability to "bend' the normal laws of nature. Since creation is thought to be the internal psychic concoction of the Cosmic Entity, it would theoretically be possible to perform unimaginable feats such as levitation if one were sufficiently advanced mentally and could somehow influence or tune the will of the Cosmic Entity to conform to their will. It is also possible that they might influence the minds of spectators, creating a collective hallucination.

Take for example the traditional Indian rope trick. In the classic version, the magician is surrounded by a crowd of people. He throws a rope into the air that is seen to disappear into the sky above. Then the magician's assistant, a boy, climbs the rope and disappears from view. The magician then calls back his assistant, but when he gets no response he becomes furious. The magician then arms himself with a knife, climbs the rope himself, and vanishes as well. An argument might be heard, and then limbs and other body parts start falling, presumably cut from the assistant by the magician. After all the parts of the body land on the ground, the magician is observed climbing down the rope. He collects the body parts and puts them in a basket. Soon the boy appears from the basket fully restored. Someone observing this scene from a distance might witness the magician sitting in meditative pose and the people around him reacting wildly. The magician is actually creating this unlikely scene in the mind of his audience by the power of his own mind, essentially hypnotizing the bystanders. Someone outside his sphere of influence sees only the magician meditating.

One problem with displaying or using psychic powers is that they may inflate the ego of the practitioner. People may be impressed by a person with such power and begin to look up to and even worship that person as a great teacher, prophet, or guru. Unless that person is fully realized like the Buddha, such admiration may lead to an inflated self-image, which is exactly the opposite of what is needed for spiritual success. Another problem with using occult powers is that they inevitably make the mind cruder and are eventually lost if used continuously. The power to heal the sick may seem beneficial; however, it can involve taking another person's karma onto oneself, which can potentially rob the sick person of a needed life lessons as well as adding that person's bad karma to the healer.

Some individuals perform spiritual practices for the sole purpose of acquiring occult powers. Others may stray from the path of knowledge and devote their energies to expanding their ego rather than minimizing it. Such individuals are termed *avidya* Tantrics in Sanskrit (the Sanskrit prefix *a* signifies negation or opposition). They can potentially do great harm to society. For example, Hitler may have been an *avidya* Tantric. He was fascinated by the occult and apparently had the power to influence people hypnotically. For this reason, spiritual teachers or gurus are very careful when accepting students and require that their pupils be well established in morality and have a yearning and love for God.

Paranormal beings

If consciousness survives the death of the body, are stories of ghosts, fairies, angels, and demons real or imagined? Since the bodiless mind does not possess a physical body, it does not possess nerves or any other faculties that would allow it to affect a physical object. This might not exclude it from affecting a person mentally. However, this too would seem unreasonable since the bodiless mind remains in a nonlocalized, dormant state. In other words, the unexpressed reactive momenta of the bodiless mind cannot find any outlet for expression until a new physical body is found for the unit.

What is behind the stories of paranormal beings? Are all such tales hoaxes or imaginary? To answer this question it is necessary to consider the meaning of what is real and what is imagined or a product of a hallucination. The seats for the sense organs are in the brain, not in the gateway organs. For example, the eyes receive light and transmit nerve impulses to the brain where the actual sensation of sight occurs. This is true of the other sense organs as well. Everything that is perceived actually occurs in the brain and is only indirectly dependent on the external world, which is what stimulates the sense organs. Therefore, nerve impulses that affect the brain but have no link to the external world via the gateway organs can appear real. Such phenomena are often labeled hallucinations, and certainly some ghost experiences fall into this category.

Hallucinations are of two kinds: positive and negative. A positive hallucination occurs when thought waves affect the sense organs in the brain and one sees, hears, feels, tastes, or smells something that is not actually present in the external world. One's sense of reality is temporarily impaired, and the conscious mind may get absorbed into the subconscious mind. Unlike the dream state, whose reality is rejected the moment one regains normal consciousness, the positive hallucination may seem real even after the conscious mind resumes normal activity. Positive hallucinations may also be elicited by hypnosis. Negative hallucinations on the other hand occur when the mind refuses to see something that is actually present. These can also be brought on by outer-suggestion (hypnosis) and by autosuggestion, most commonly because of fear, when one's mind refuses to accept some aspect of an experience.

Fear has a powerful effect on the mind. It can cause temporary concentration. For example, if a person believes in the existence of ghosts and visits a house that is said to be haunted, fear may trigger a positive hallucination of a ghost. Theoretically, concentration of mind caused by fear can leave an imprint in the mind-stuff at that location. That in turn may trigger the subconscious minds of other people and cause them to experience similar hallucinations. Therefore, most ghost stories are propagated by fear and have no physical reality.

However, if the psyche of a person is disturbed enough it potentially could move physical bodies. Stories of entities moving physical objects may be due to the action of mind-stuff on physical matter. However, such activity is not the result of the action of a bodiless mind and is wholly dependent on the mental action of a living human being.

Sages have identified a type of paranormal being, which they call a *devayoni*. *Devayoni* means "luminous body," and these beings may be responsible for some of the stories about angels and other luminous beings. They are believed to be advanced souls who were unable to attain liberation due to some desire or attachment that remained in their mind at the time of death. For example, a person may have meditated regularly for most of their life and become very subtle and spiritually advanced. Perhaps this person had a great love of the fine arts, especially the intoxicating rhythms of fine music. At the time of death such a person may lack the necessary devotion for complete surrender of their individuality or have a desire to continue to enjoy music. If they are unable to surrender their unit consciousness completely and merge with Cosmic Consciousness at the time of death, then their mind may be too subtle to be reborn directly in a normal physical body consisting of the five fundamental factors. Instead, they take on a body having only ethereal, aerial, and luminous factors. Lacking a complete physical body with liquid and solid factors, they will be unable to perform meditation and continue in a normal manner on the path to liberation. They will be able to enjoy music sympathetically by frequenting concert halls and other places where music is played. However, until their desire for music is satiated and all those reactive momenta burned, they will remain as a luminous body and will not be able to be reborn in a physical body.

Sages have identified seven different types of *devayonis*, according to their different desires. The most highly developed is the siddha. A siddha is a spiritually advanced soul who had a desire at the time of death to be reborn with spiritual powers and do great work. Such beings enjoy sympathetically the vibrations emanated by persons doing meditation and by persons who are highly advanced spiritually. Since *devayonis* have the luminous factor, they may at times be perceived with the naked eye. Their presence may also be felt under the right circumstances. Many sightings of angels, fairies, and friendly ghosts may in fact be luminous bodies.

Summary

The existence of a psychic body that parallels the physical body explains how a complex organism consisting of trillions of individual cells can function as a single integrated living organism. The psychic body is like the conductor of a symphony orchestrating the nerve and chemical messages that the physical body of an organism requires so that it can function as an integrated unit.

Psychic powers and ESP do exist. Mind is nonlocal and the unconscious layers of mind are universal in character. Individuals are sometimes capable of tapping this reservoir of information to perceive knowledge of future events or things beyond the ordinary senses. Extramundane knowledge can be obtained by persons trained in the art or be demonstrated at a less robust level in a large number of people using statistics.

Mental intent can cause the wave function of the brain or a physical object (e.g. a RNG) to collapse in a specific way and thus move the body or possibly affect a physical object.

12
Explaining Scientific Anomalies

THE SPIRITUAL WORLDVIEW OF how the universe evolved is mostly in accord with that of modern cosmology. However, all scientific theories of creation rest on the assumption that something was present or came into existence at the time of creation. But where did this something come from and how did it become so organized? In other words, what is the first cause of creation and why are the physical laws guiding creation tuned for human existence? Naturally, scientists and cosmologists that study the origins and nature of the universe try to avoid such questions by saying that these ultimate questions are the purview of metaphysics or religion, not astrophysics. More and more scientists, however, have turned to the study of Consciousness as an explanation for the origin of the universe. The *Brahma Chakra* theory of creation as described in Chapter 10 provides an explanation for how the universe and life originated from Consciousness. Consciousness is the first and last cause. The improbability that life originated from dead matter is explained by the fact that matter is ultimately composed of Consciousness and therefore has the potential of expressing life. Hence to call it "dead matter" is incorrect. This theory answers the ultimate questions about the origin and evolution of the cosmos and living organisms in an elegant and logical way. It also avoids the illogical proposition described in the Bible that the first cause (God) is separate from his creation.

The evolution of species was first theorized by Darwin. It has been refined by our understanding of heredity and the role of DNA. Today the orthodox view of evolution is that it can be reduced to physio-chemical processes that are passed on genetically to organisms that adapt best to their environment. The reductionist doctrine of evolution is now accepted by nearly all biologists and one of its best spokesman is Richard Dawkins, who wrote the book *The Blind Watchmaker: Why the Evidence of Evolution Reveals a Universe without Design.*[1] We can call the current theory of evolution neo- or material-Darwinism. There is no need to postulate an organizing principle; random events and the passage of time can explain the evolution of species.

There is no doubt that Darwin's theory is largely correct. The evolution of species takes place because of natural selection and survival of the fittest. Small changes in the genetic material of an organism over a long period can lead to major changes in that organism and possibly the development of new species. It is now well established that changes in DNA, which normally involve random single-point mutations, can sometimes promote the survival of an organism. For example, bacteria become resistant to an antibiotic by the mutation of a single gene, and peppered moths became darker due to industrial pollution during the Industrial Revolution. While evidence clearly shows that evolution can produce changes within a species, there is less evidence to support the theory that it can generate the much greater diversity that exists between species.

It turns out that the confidence that modern science has in reductionist neo-Darwinism is not based on scientific evidence but on assumptions based on predisposed attitudes similar to what we see in the neurosciences. As long as there is a credible argument based on purely physio-chemical mechanisms, there is no reason to think outside the box. However, this simplistic argument does not hold up when the evolution of complex systems, structures, mind, and consciousness is considered.

One example of a complex system that cannot be reduced to individual parts is blood-clotting in mammals. No less than eleven enzymes are involved in this intricate cascading system with several feedback loops. If any one of the enzymes is missing or defective, it will be a death sentence for the organism. Thus hemophiliacs, who have one defective gene and are missing a single blood-clotting enzyme, inevitably die young unless they receive modern treatments. Such an irreducibly complex system requires that essentially all the components are present before there is any advantage conferred to the organism.[2] Hence, it is very hard to understand how such a system could evolve in a step-wise manner via random mutations of DNA with selection of the fittest. The odds against all the required components of the system arising simultaneously are enormous. In addition to blood-clotting, there are other systems and structures in living creatures that must be considered irreducibly complex. Examples include the bacterial flagellum, the eye, and the immune system.

The materialist approach to evolution fails to account for how mind and consciousness evolved. There is no plausible physicalist theory to account for consciousness. Those in the reductionist camp explain that this does not mean they will not discover a material basis for consciousness in the future. After all, they say, it has taken science many years to discover the basis for many natural phenomena; they are working on this problem and have made some progress. However, the problem would seem to be insurmountable. As was pointed out before, a viable brain-based theory of mind cannot be shown to exist, so how can consciousness be reduced to matter? Philosopher Thomas Nagel makes a strong argument that since mind and consciousness are essential features of living organisms, neo-Darwinism needs to explain how they evolved and why. It is not enough to simply say they came into existence because of physical changes in the organism. Any account of biologic evolution must explain the appearance of conscious organisms.[3] In other words, the materialist argument that mind "emerges" after a certain level of complexity arises in a living organism sidesteps the question of where it comes from. For mind to arise from matter implies that it exists in a potential form in that matter—exactly the argument of spirituality.

Another problem with neo-Darwinism is the existence of gaps in the fossil record. According to the theory, evolution should be slow and continuous. However, the existence of gaps in the record suggest that there has been discontinuity or unexplained "jumps" in the evolution of species.[4] One example of this was the explosion of fauna, seemingly abruptly and from nowhere, that occurred during the early Cambrian Period. Darwin recognized that this relatively short evolutionary event (20-25 million years) might be one of the main objections that could be made against his theory of evolution by natural selection.

The most logical conclusion is that the evolution of species is teleological and that mind and consciousness are the basic ground substances that underpin life. In other words, evolution depends on the design or influence of Cosmic Mind. This explains how life originated from dead matter, how distinct species evolved rather rapidly, and how complex physical systems and structures evolved.

Animal instincts

The modern science of genetics has made great strides in explaining how the physical attributes of living organisms are passed on between generations.

However, genetics has yet to elucidate how complex instinctual behavior in animals is passed from one generation to the other. DNA is the genetic material responsible for the transmission of the physical structures (cells) of all living organisms (with the exception of some viruses that utilize RNA). It is the blueprint for constructing the entire organism starting from a single cell. DNA is nature's code for synthesizing proteins from constituent amino acids, and proteins are the basic building blocks of living tissues. They play a central role in biological processes. For example, enzymatic proteins catalyze chemical reactions and keep the machinery of life going. Others transport oxygen, move joints, run the immune system, and carry messages from cell to cell.

Many instincts in animals can be traced to how the brain and nervous system develop, and to glands and certain structures and neural networks that affect specific behaviors or cause the organism to respond to stimuli under specific circumstances. Genetics might explain simple autonomic responses and behaviors that promote self-preservation, but complex instinctual behavior is not so easily understood. For example, a bowerbird has an extraordinarily complex courtship and mating dance and the males build a complex nest (bower) to attract mates. This is not learned behavior. In order for the behavior to be inherited, it would have to be hardwired or programed into the bird's DNA. This would be tantamount to hypothesizing that there exists a genetic form of memory. The problem is that there is no evidence directly linking DNA with memory. If there were genetic memory, then it would be reasonable to expect that we would inherit knowledge from our parents, grandparents, etc. Yet such knowledge is lacking and there is no known repository of inherited knowledge in either the genome of humans or in animals.

Many other animals show extremely complex instinctive behavior. Even spiders spin webs of a specific and complex design that is not learned from a parent. How and why do the spiders produce a web of exactly the same design? One explanation is that such unlearned behavior is passed from one generation to the next by a complex yet unknown mechanism involving DNA, proteins, or RNA. A simple explanation is that many animal instincts are a result of the reflection of Cosmic Mind on the rudimentary mind of the organism.

Certain examples of homing behavior in animals are probably due to the connection animals have with Cosmic Mind. Olfactory memory has been suggested as the mechanism by which salmon are able to return to the stream they were born in to spawn. However, for Atlantic salmon this would require that the "smell" of a stream not change significantly over a period of thirteen years; otherwise most fish would stray, which is not observed. As mentioned earlier, the homing and remarkably precise long-distance migrations of birds, monarch butterflies, turtles, etc. is still not understood by scientists. They have suggested that homing and migrating animals use navigational clues such as the sun angle, star patterns, and Earth's magnetic field, but homing and migrations may occur in any compass direction and during any season. The alternative explanation is that they utilize nonlocal mind.

Fine-tuning

Today scientists recognize that the cosmological constant, fundamental forces, physical laws and constants all seem to be exquisitely fine-tuned to allow for the existence of stars, planets, water, and ultimately for the emergence of living organisms and self-conscious beings. In Chapter 1, it was noted that the expansion of space-time (cosmological constant or lambda) was exactly balanced by the total mass-energy of the universe.

Noted physicist Stephen Hawking had this to say about this enigma:

> Why is the universe so close to the dividing line between collapsing again and expanding indefinitely? In order to be as close as we are now, the rate of expansion early on had to be chosen fantastically accurately. If the rate of expansion one second after the Big Bang had been less by one part in ten billion, the universe would have collapsed after a few million years. If it had been greater by one part in ten billion, the universe would have been essentially empty after a few million years. In neither case would it have lasted long enough for life to develop. Thus one either has to appeal to the anthropic principle or find some physical explanation of why the universe is the way it is.[5]

There are four fundamental forces responsible for shaping the universe. These are gravity, electromagnetic, weak and strong nuclear forces. The distances covered and strengths of the four forces differ greatly. The strong nuclear force is extremely strong since it must overcome the mutual repulsion of protons in the nucleus of atoms; however, it only acts over an extremely short distance. The electromagnetic and weak forces in atoms are intermediate in strength and act over a short distance (for example, the electromagnetic attraction of electrons for protons). However, the electromagnetic force may cover great distances as radiation, such as light. Gravity is an extremely weak force but it can act over astronomical distances.

The exquisite fine-tuning of these four forces is required for a long-lived universe and for the conditions necessary for the emergence of life on Earth. For example, the strong nuclear force, which binds protons and neutrons in the nucleus of atoms, is just the right strength to make stable atomic nuclei and allow the fusion of hydrogen nuclei in the center of stars. This releases just the right amount of energy to power stars for a very long time, thus maintaining a more moderate and constant temperature on planets such as Earth. This enables higher life forms to evolve, a process that can take billions of years. Similarly, the weak force is perfectly tuned to allow for the nuclear decay of heavy elements, which heats Earth's molten core, resulting in volcanism and a magnetic field that shields life on Earth from deadly cosmic radiation.

The electromagnetic force is responsible for light, heat, electricity, and for the chemistry that is responsible for matter as we know it. Even a slight variation in its strength would have a profound effect upon the chemistry of life. For example, the unique properties of water, such as its high freezing and boiling points, solvating power, and expansion upon freezing are all due to its strong polarity or asymmetric charge distribution. These properties have been essential for the evolution of life on Earth. If the universal electromagnetic forces at work holding molecules together and the similar forces responsible for creating polarity within molecules were even slightly different, then it is unlikely that water would have the unique properties needed to support life, nor would proteins or DNA have the same properties.

If the force of gravity differed even slightly from its actual value, models predict that stars, galaxies, planets, and hence living organisms could not have developed from the dust left by the primordial Big Bang. Even the formation of matter following the Big Bang depended on a tiny excess of matter particles over antimatter particles—something that would not be predicted by current theories. The universe as we know it could not exist without a slight excess of matter particles over antimatter particles following the Big Bang.

Physicist Roger Penrose noted that the fine-tuning necessary for life to emerge from the Big Bang in terms of phase-space-volume relationships is so absurdly unlikely to have occurred by chance that there must be another explanation.[7] Sir Martin Rees of the Royal Observatory in Britain wrote that there are just six numbers necessary for our emergence from the Big Bang as conscious beings. However, if there were a minuscule change in any one of these cosmological constants then life and the universe as we know them would be impossible.[8]

Many other scientists have pointed out how incredibly fortunate we are to have a universe that allows for the development of living organisms. The explanation provided by the materialist camp is called the anthropic principle—only in a universe capable of eventually supporting life will there be living beings capable of observing and reflecting upon fine-tuning. At the heart of this idea is the allowance for multiple universes—multiverse theory. According to multiverse theory, during the chaotic inflation phase immediately following the Big Bang an untold number of universes were created, each with slightly different rates of expansion, cosmological constants, and forces. Out of the almost infinite universes created, ours has just the right conditions for the development of conscious beings.

The Standard Model of particle physics

In order to understand if multiverse theory is a viable explanation for fine-tuning, it is advantageous to touch on the current model used to describe the forces that shape the universe. It is known that all four forces act at a distance and are therefore field forces. That means that the greater the distance between objects the weaker the force.

The best theory that unifies these forces, with the exception of gravity, along with all the subatomic particles observed is called the Standard Model of particle physics. This theory proposes that the known particles are made from smaller particles called quarks. The existence of the six quarks has been experimentally verified. The theory also describes force fields as utilizing intermediate particles (gauge bosons) as carriers of the forces. For example, the electromagnetic attraction between an electron and proton is carried by the photon, while the strong nuclear force is carried by the gluon and the gravitational force by the graviton.

The one missing particle in the theory, which was required if some fundamental particles were to have mass, was the Higgs boson. In 2012 the existence of the Higgs boson was demonstrated using the Large Hadron Collider in Cern, Switzerland. This was a landmark accomplishment and served to cement the Standard Model. Although the Standard Model is theoretically self-consistent and has demonstrated great and continued successes in providing experimental predictions, it falls short of being a theory of everything because it fails to provide a theory for gravity. The Standard Model has been very successful in providing a fundamental understanding of the interactions and composition of matter and energy at the quantum level, and successful in predicting and describing with great accuracy experimental results over an enormous range of energies.

A new hypothetical model that goes beyond the Standard Model is called supersymmetry. It is an extension to the Standard Model that aims to fill some of the gaps. It predicts a partner particle for each particle in the Standard Model whose spin differs by one-half. These new particles would solve a major problem with the Standard Model—fixing the mass of the Higgs boson and insuring that it would be stable. Since the Higgs holds everything together, if it were unstable then the universe would eventually dissolve. The supersymmetry model predicted that the Higgs should have a mass of about 115 GeV. On the other hand, multiverse theory predicted a mass of approximately 140 GeV for the Higgs. Cern scientists found that the actual mass of the Higgs was 126 GeV, intermediate between the two but failing to provide any evidence in support of multiverse theory. Hence, this explanation for fine-tuning remains purely speculative, probably untestable, and is highly complex. Supersymmetry theory will undoubtedly win the day because it is consistent with spiritual ideology in proposing that underlying this seemingly complex relative reality lies unity. Supersymmetry also satisfies Ockham's razor, which states that when there is more than one competing hypothesis to explain a phenomenon, both of which are equally plausible, it is the simplest explanation that is normally correct. Eventually scientists will discover a mathematical model that explains how the forces and particles found in nature are interconnected, thus reflecting the wholeness that is fundamental to the universe.

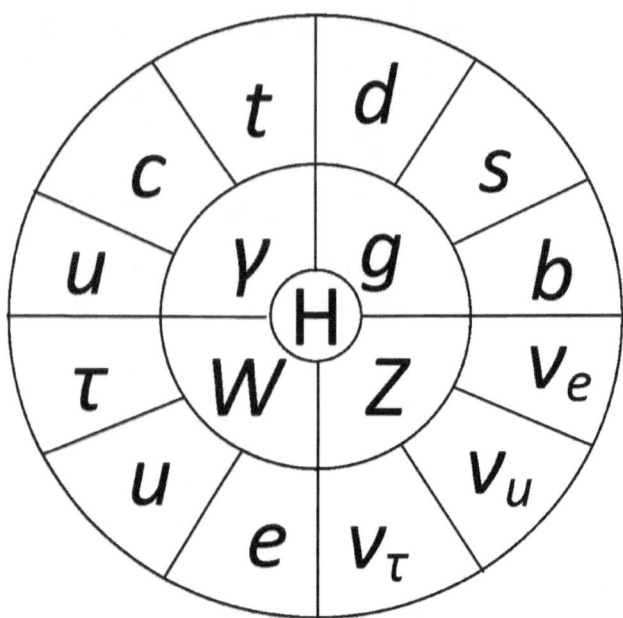

Diagram of the Standard Model. The Higgs boson (H) is centrally located and is the lynch pin of the model. The middle circle shows the gauge particles for the four forces (γ-photon, g-gluon, W-boson, Z-boson). The outer circle has the six quarks on top (u-up, c-charm, t-top, d-down, s-strange, b-bottom). The bottom six are the leptons (τ-tau, u-muon, e-electron, V_τ-tau neutrino, V_u-muon neutrino V_e-electron neutrino).

Domain of the wave function

Physics has shown that there exists a hidden domain that lies at a level of reality that is subtler than what is experienced with the mind and sense organs. This may be termed the "domain of the wave function." Some of the properties of this domain are the following:

- *It expresses wholeness.* This is a fundamental aspect of reality at a level deeper than ordinary space-time.
- *It is timeless.* In exists outside space-time and the past, present, and future are meaningless when discussing this realm. It is only after the wave function collapses by observation that an arrow for time comes into existence.
- *It is nonlocal.* It penetrates and surrounds ordinary reality and is not localized in any part of space but is all encompassing, everywhere at the same time. It is only after observation that a "part" of reality represented by the function becomes localized in space.
- *It is a mathematical representation of the possibilities that reality may take.* In addition, it determines the probability that any particular possibility will become reality.
- *It is connected to everything.* Parts of the whole emerge from this realm when consciousness is directed or focused on one of the potential states allowed by the function.
- *All matter and energy have associated wave functions.* This includes the brain, the body, and the universe as a whole. The wave function represents the gestalts for these individual entities.
- *The domain of the wave function does not contain energy as such.* Instead, it can be considered the underlying source of all energy, and therefore, in a sense contains the potential for expression of almost infinite energy.

It should be obvious that the properties of the wave function are the same as those of Cosmic Mind. Hence, physics describes a subtle, hidden realm that has identical properties to Cosmic Mind, and this realm must function when individuals express any thought, action, desire, etc. Nothing could take place in one's mind without some form of conscious intent, just as it is impossible for any particle of matter to come into existence without an act of observation.[9] Hence, physics validates both the existence of Cosmic Mind and the role that the "I do" function of mind plays in creating "reality." However, the reality that is perceived must be considered a relative reality because it originates from a subtler, timeless, all-encompassing realm of the wave function, which corresponds to the all-knowing mind of God.

Quantum mechanics tells us that not all possibilities described by the wave function have equal probability of becoming manifest when there is observation. Hence, it seems to describe a reality that is probabilistic and uncertain (think Heisenberg uncertainty principle). However, if the universe is created from Cosmic Consciousness and is transformed into Cosmic Mind, which permeates but transcends space-time, then in reality there could be no uncertainty. How can these two seemingly incompatible aspects of reality be rationalized? Clearly, uncertainty in measuring quanta is a result of the wave nature of matter, energy, and mind. Mind is nonlocal, energy is nonlocal, and matter is more localized but still cannot be pinned down exactly in space because it too has wave nature. All measurements are performed with devices, which are also nonlocal, and consequently there is some uncertainty in any measurement.

Physics teaches that probability applies when it comes to which of the possible states of a wave function will emerge into reality upon conscious observation. Does this mean that the universe is a chaotic system in which any number of events can happen merely by random chance? The answer is a resounding no! Theoretically, the whole universe has a wave function, which could also be called Cosmic Mind. Every particle, every quanta of energy, every event that has or will take place happens for a purpose; nothing is by chance—everything is interconnected. The entire creation is nothing but the internal mental concoction of the Cosmic Entity. Events may appear random or coincidental to the casual observer but only because one is unable to see the big picture.

Summary

The materialist worldview teaches that our universe is one of countless universes created in a mysterious process in which multiple universes were simultaneously created from nothingness like a very long sheet of bubble wrap. Each of the universes differed in their total mass, rates of expansion, and physical forces. In our very "special" universe, all these conditions were just right so that conscious life could evolve. But like the bubbles in the wrap, each universe is separate and there is no way to test this hypothesis.

The materialist worldview requires that against enormous odds the chemical precursors for life, such as amino acids and nucleotides came together by random chemical reactions to form the genetic material and all the machinery required for reproduction and the metabolic processes required for life. Furthermore, all the diversity, complexity, instinctual behavior, and development of mind and consciousness were also the product of random alterations of an organism's genetic material (DNA) concomitant with the selection of advantageous mutations passed on to offspring over a long time.

There is little logic in embracing this worldview, and those proponents of it appear to have decided that mind and consciousness are merely products of matter and any evidence to the contrary must be false.

On the other hand, the spiritual worldview is self-consistent, simple, and logical. Furthermore, it fits the facts and it explains both physical and natural phenomena that the alternative worldview cannot. It has only one basic assumption—that Consciousness is not an epiphenomenon of matter but the ground substance of creation—the beginning and the end. The universe and everything in it has evolved from Consciousness and life is a natural expression of the tendency of Consciousness to be transformed into individual life forms that are inexorably drawn by attraction for the Great back toward the Source.

13
Science and Metaphysics

METAPHYSICS IS A DIVISION of philosophy that is concerned with the fundamental nature of reality and being. One of the principle branches of metaphysics is cosmology, which is the study of the origin, fundamental structure, nature, and dynamics of the universe. Metaphysics as it pertains to the spiritual worldview is the philosophical statement that the universe is an indivisible whole, within which constituent parts exist at a "lower" level of reality and are connected with each other via their connection with the whole. This philosophical theory can be called "holism," and it tries to address the big questions concerning existence, the nature of reality, and God. Science is also interested in the nature of reality but is an intellectual and practical activity that utilizes observation and experimentation to elucidate the structure and behavior of the physical and natural world. These two systems of discovery would seem to be incompatible. Metaphysics is philosophy and it is primarily based on intellectual ideas and subjective experience. On the other hand, science is based on objective experience and observation.

In its early history science was inexorably entwined with metaphysics. For example, alchemists were early scientists who attempted to transmute base metals such as lead into gold. Less known is that they also aimed to create elixirs of immortality and panaceas able to cure any disease, and that their ultimate goal was the transformation of the soul to achieve knowledge and power.

This early alliance of science and metaphysics began to wane in the late seventeenth and eighteenth centuries with the discoveries of Galileo and Newton and with the philosophy of such men as René Descartes. The latter argued that there was a clear distinction between mind (the purview of metaphysics or religion) and matter, which fell into the domain of science. During the nineteenth and early twentieth centuries, the scientific pendulum swung to the opposite extreme as it began to divorce itself completely from metaphysics. Scientists of the time believed that all natural phenomena including how life evolved could be explained by natural laws that governed the universe. There was no longer any need to attribute creation to a supernatural being or God.

However, this separation of science and metaphysics changed in the early part of the twentieth century with the discovery of quantum physics. Some of the early pioneers in this new scientific description of reality immediately recognized that it supported an ontology where mind and consciousness take center stage at the expense of matter. Metaphysics again entered the realm of science and was important for understanding the results of scientific observations.

One of the first to recognize this was Max Planck (1858–1947). He was truly the father of quantum physics. Born and raised in Germany, he was the first scientist to postulate that nature is not continuous but consists of discrete packets of energy or quanta. Based on his studies of the radiation emitted by a heated piece of metal (known as black-body radiation), he postulated that the electromagnetic radiation emitted by the body could only be a multiple of a constant (Planck's constant) times the frequency of the radiation. Planck's discovery ushered in the new era of quantum mechanics, and he may have been one of the first scientists to see the connection between the new physics of the quantum realm and its interplay with Consciousness and the unity of all things. He is quoted as saying:

> I regard consciousness as fundamental. I regard matter as a derivative of consciousness. We cannot get behind consciousness. Everything we talk about, everything that we postulate as existing, requires consciousness.[1]

Another of the pioneers of quantum physics was Wolfgang Pauli (1900–1958). He was born in Austria and was a brilliant scientist who is best known for his discovery of spin theory and the Pauli exclusion principle that bears his name. This principle was important for understanding the structure of matter, and chemistry in particular.

Pauli was not your average empirical scientist. He believed that the universe had inherent order and that it could best be understood using intuition, or through experience of certain archetypes that exist in the collective unconscious. In this respect, his beliefs were similar to those of the seventeenth-century mathematician and astronomer Johannes Kepler and the modern psychologist Carl Jung.

Pauli had an understanding of the unity and connection of all things that was reflected in the complementary aspects of physical phenomena.

> The idea of complementarity in modern physics has demonstrated to us, in a new kind of synthesis, that the contradiction in the applications of old contrasting conceptions (such as particle and wave) is only apparent... The only acceptable point of view appears to be the one that recognizes both sides of reality—the quantitative and the qualitative, the physical and the psychical—as compatible with each other and can embrace them simultaneously.[2]

Pauli called for the use of metaphysics for obtaining an understanding of the "deeper invisible reality" that transcends the traditional macroscopic descriptions of reality.

> For I suspect that the alchemist's attempt at a unitary psychophysical language miscarried only because it was related to a visible concrete reality. But, in physics today, we have an invisible reality (of atomic objects) in which the observer intervenes with a certain freedom (and is thereby confronted with the alternatives of "choice and sacrifice"); in the psychology of the unconscious we have processes which cannot always be unambiguously ascribed to a particular subject. The attempt at a psychophysical monism seems to me now essentially more promising, given that the relevant unitary language (unknown as yet, and neutral in regard to the psychophysical antithesis) would relate to a deeper invisible reality. We should then have found a mode of expression for the unity of all being, transcending the causality of classical physics as a form of correspondence; a unity of which the psychophysical interrelation, and the coincidence of a priori instinctive forms of ideation with external perceptions, are special cases. On such view, traditional ontology and metaphysics become the sacrifice, but the choice falls on the unity of being.[3]

Pauli also commented on the mystical unity of mind, body, and spirit in a letter to theologians:

> I shall not venture to make predictions about the future. But, contrary to the strict division of the activity of the human spirit

into separate departments—a division prevailing since the nineteenth century—I consider the ambition of overcoming opposites, including also a synthesis embracing both rational understanding and the mystical experience of unity, to be the mythos, spoken or unspoken, of our present day and age.[4]

Sir James Jeans (1877–1946) was another scientist that made many important contributions to quantum physics, the dynamic theory of gases, radiation, and stellar evolution. In his book *The Mysterious Universe*, he outlined his belief that everything in the universe is entangled.

> It is the same, I think, with other more technical concepts, typified by the "exclusion principle," which seem to imply a sort of "action-at-a-distance" in both space and time—as though every bit of the universe knew what other distant bits were doing, and acted accordingly. To my mind, the laws which nature obeys are less suggestive of those, which a machine obeys in its motion than those, which a musician obeys in writing a fugue or a poet in composing a sonnet. The motions of electrons and atoms do not resemble those of the parts of a locomotive so much as those of the dancers in cotillion.[5]

He also believed that the universe had to be the internal psychic thought projection of the Cosmic Entity:

> If the universe is a universe of thought, then its creation must have been an act of thought. Indeed, the finiteness of time and space almost compel us, of themselves, to picture the creation as an act of thought; the determination of the constants such as the radius of the universe and the number of electrons it contained imply thought, whose richness is measured by the immensity of these quantities. Time and space, which form the setting for the thought, must have come into being as part of this act. Primitive cosmologies pictured a creator working in space and time, forging sun, moon, and stars out of already existent raw material. Modern scientific theory compels us to think of the creator as working outside time and space—which are part of his creation—just as the artist is outside his canvas. It accords with the conjecture of Augustine: *Non in tempore, se cum tempore, finxit Deus mundum*

(God creates the universe not in time but along with time). Indeed, the doctrine dates back as far as Plato (time and the heavens came into being at the same instant…Such was the mind and thought of God in the creation of time).[6]

Jeans disagreed with the philosophy of Descartes that mind and matter were separate and not connected. If this were true then the mind could not possibly affect the body. Thus Jeans agreed with idealist philosophers such as George Berkeley who argued that since mind and matter do in fact interact, they must be of the same nature and that matter must arise from mind.

Another important personality in the development of quantum mechanics was Erwin Schrödinger (1887–1961), who is best known for his pioneering work developing the quantum wave equation that bears his name and for which he received the Nobel Prize in Physics in 1933. Similar to Carl Jung and his realization that there exists a collective unconscious, Schrödinger believed in what he called the "One Mind" and in the unity of all things. Schrödinger held that this idea was supported by science. Schrödinger had a keen mystical insight into the functioning of the universe. Examples of his ideas on wholeness are taken from his books *My View of the World*, *Mind and Matter* and *What Is Life?*

> Subject and object are only one. The barrier between them cannot be said to have broken down as a result of recent experience in the physical sciences, for this barrier does not exist.[7]

He considered individual consciousness to be a reflection of a higher Self connecting individuals with one another and with the universe.

> To divide or multiply consciousness is something meaningless. In all the world, there is no kind of framework within which we can find consciousness in the plural; this is simply something we construct because of the spacio-temporal plurality of individuals, but it is a false construction. The categories of number, of whole and of parts are then simply not applicable to it; the most adequate expression of the situation being this: the self-consciousness of the individual members are numerically identical both with one another and with that Self which they may be said to form at a higher level.[8]

His comments on mystical union are a statement of the spiritual worldview.

> There is obviously only one alternative, namely the unification of minds or consciousness. Their multiplicity is only apparent; in truth, there is only one mind. This is the doctrine of the Upanishads, and not only of the Upanishads. The mystically experienced union with God regularly entails this attitude unless it is opposed by strong existing prejudices; this means that it is less easily accepted in the West than in the East.[9]

Schrödinger argued that our "I" is the "I" of God. He pointed out that our body functions as a pure mechanism according to the laws of nature; yet we know by incontrovertible direct experience that we are directing its actions. Hence he argued that the "I" that controls the motion of atoms must be the "I" of God. Although this metaphysical statement is strange by Western modes of thinking, he wrote that it is seen consistently in the writing of mystics and is essential to Eastern and spiritual philosophy.

Werner Heisenberg (1901–1976) is best known for his uncertainty principle, but he contributed much to quantum physics and received the Nobel Prize in Physics in 1932. His understanding of quantum physics led him to express in several of his writings that separateness or the existence of distinct and nameable parts may be meaningless for the ultimate reality that physics seeks to deal with. For example, in his book *Across the Frontiers* he wrote:

> We know that there is an ever-changing variety of phenomena appearing to our senses. Yet we believe that ultimately it should be possible to trace them back somehow to some one principle.[10]

He believed that the smallest units of matter are not physical objects in the ordinary sense of the word but forms that arise from a world of potentialities or possibilities rather than one of things, and that the whole is formed from the intermingling of the parts.[11]

David Bohm (1917–1992) was born in the United States and made many important contributions to theoretical physics. In his book *Wholeness and the Implicate Order*, he argued extensively that modern scientific theory implies that the universe displays indivisible wholeness.

> Ultimately, the entire universe (with all its particles, including those constituting human beings, their laboratories, observing

instruments, etc.) has to be understood as a single undivided whole, in which analysis into separately and independently existent parts has no fundamental status.[12]

Bohm realized that underpinning everyday physical reality was a deeper and far more pervasive reality.

> Quantum mechanics suggests that this is the way that phenomenal reality comes about from a deeper order in which it is enfolded. Reality unfolds to produce the visible order and folds back in. It is constantly unfolding and enfolding.[13]

Bohm recognized that quantum physics was only compatible with a spiritual worldview. He wrote:

> Deep down the consciousness of mankind is one. This is a virtual certainty because even in the vacuum matter is one; and if we don't see this it's because we are blinding ourselves to it.[14]

Another scientist turned philosopher was Henry Margenau (1901–1997). He was a German-born, US-educated physicist whose greatest contribution was not in the field of quantum physics but in integrating science and religion. His most important work in this field was his book *The Miracle of Existence*. Margenau argued that science as well as simple observation points to an undivided wholeness in the universe, which he called "Universal Mind." He argued that the concept of separateness or discreteness does not apply to the quantum level for matter and similarly cannot be applied to the mental or superconscious levels. A mind without constituent parts is his Universal Mind, which is similar to the collective mind of Jung and Schrödinger and to Cosmic Mind. Margenau argued that the fact that different living entities all perceive the same world despite differences in their brains is evidence for Universal Mind. He argued that considering the fact that the world is known only through the sense organs and the brain, both of which differ greatly between individuals, it is remarkable that everyone perceives the same picture of the world. This is only possible because everyone shares the same One Mind. In other words, it is only because all people share the same Consciousness that they perceive things the same way. If this were not true then there would be many different perceptions of reality. He wrote:

If my conclusions are correct, each individual is part of God or part of the Universal Mind. I use the phrase "part of" with hesitation, recalling its looseness and inapplicability even in recent physics. Perhaps a better way to put the matter is to say that each of us is the Universal Mind but inflicted with limitations that obscure all but a tiny fraction of its aspects and properties.[15]

Margenau believed that mind was nonmaterial and could function separately from the brain but was capable of influencing the brain on the quantum level, which would not require the mind to expend any energy in the process. He pointed out that conservation of energy as normally understood does not always apply on the quantum level—for example electrons can pass through barriers without expending any energy and particles with mass can be created out of the void of space. He was perhaps the first to point out that in mind-body interactions, the brain can supply any energy needed as mind causes a collapse of the wave function resulting in a particular probability event to take place in the physical organ.

If each individual mind is merely an illusory part of Universal Mind, then why do people feel separate and bound by time, place, and person? Margenau would argue that it is their ego or personal wall, built up from childhood, that reinforces the separateness they feel. He believed that these limitations are overcome by the mystic who feels "at one" with the Cosmic Entity.

Finally we come to one of the greatest minds of the twentieth century, Albert Einstein (1875–1955). He is best known for his work on the theory of relativity and the nature of space-time, gravity, and equivalence of mass and energy. However, Einstein also made major contributions in the realm of quantum physics and received the Nobel Prize in 1921 for his work on the photoelectric effect—a quantum phenomenon.

For most of his life, Einstein subscribed to the materialist worldview and believed that there should be a one-to-one correspondence between physical reality and physical theory to explain that reality. Perhaps because of this bias and because he believed that nothing in the physical universe could exceed the speed of light, he had a hard time accepting the quantum physics of Neils Bohr and his colleagues in Denmark (the Copenhagen Interpretation). The theory postulated that entangled quantum particles were in instantaneous or superluminal communication with one another, no matter how far apart they were. Furthermore, the idea that there is uncertainty in every quantum event went against his sentiment that if there were a God, he would not create a universe governed by chance occurrences.

Although Einstein came up with many scientific objections to quantum theory, ultimately his objections were shown to be without merit and the physics of Bohr and colleagues was proven correct. Near the end of his life, Einstein apparently realized that the nonlocality of space-time (his hypothesis) and quanta (Bohr's hypothesis) implied that there was a fundamental unity of all things, for he is quoted as saying:

> A human being is a part of the Whole, called by us the "Universe," a part limited in time and space. He experiences himself, his thoughts and feelings as something separate from the rest—a kind of optical illusion of his consciousness. This delusion is a kind of prison for us, restricting us to our personal desires and to affection for a few persons nearest to us. Our task must be to free ourselves from the prison by widening our circle of compassion to embrace all living creatures and the whole of nature in its beauty. Nobody is able to achieve this completely, but the striving for such achievement is in itself a part of the liberation and a foundation for inner security.[16]

In this quote, Einstein is exclaiming the basic concept of spiritual ideology that discreteness is a macroscopic illusion. Einstein also addresses liberation of the self or freeing oneself from the illusory bondage of the personal ego:

> The true value of a human being is determined primarily by the measure and the sense in which he has attained to liberation from the self.[17]

These great physicists from the past saw compelling experimental evidence for the essential role of mind in affecting physical events and realized that it was no longer possible to think in terms of distinct parts when it came to the quantum level. Therefore, if distinct parts cannot be applied to the physical realm then how can individual parts apply to mind? Mind without parts is the One Mind, Universal Mind, Cosmic Mind, Consciousness, Brahma, God, or any number of other names. These scientists had the keen insight to realize that the discoveries of quantum physics and relativity theory paved the way to a new scientific paradigm based on the idea that Consciousness is the ground substance of creation.

Science in metaphysics

Modern science needs to include Consciousness in its repertoire of ideas if it is to progress beyond the narrow materialistic worldview. However, it is equally important that modern-day metaphysics reflect the discoveries of science. Here we use the term metaphysics as the philosophical foundation behind theological (religious), mystical, and spiritual ideologies. In the past, most of the tension in this arena has been between science and religion. When religious doctrine contradicted the discoveries of science, theologians were often slow to accept the scientific version of reality. Examples include the Catholic Church's resistance to a heliocentric solar system and to Darwin's theory of the evolution of species. Although some religious fundamentalists might still reject the theory of evolution in favor of creationism, today scientific discoveries are accepted as facts of nature by most theologians.

A bigger problem with Western theology might be the problem of dualism. Is it logical that an all-encompassing Creator could be separate from his creation? The traditional theism of the Abrahamic religions of Judaism, Christianity, and Islam describe a God that is above and beyond his creation. This tenet of classical theism describes God as the cause of the world, but he is not the material cause of the world. That is, the stuff of the creation is different from God, who is totally immaterial and perfect (free of sin). God is spirit and the world is matter. Hence, classic theism maintains that God is present and active in governing and organizing the universe but not "in" his creation except as spirit. This is required by the doctrine that God is perfect, yet his creation is imperfect. The problem with this view of God is that it assumes that God can create something separate from himself, i.e. the material universe. In other words, God is separate from his creation and it is not contained within the Divine Being. If he is truly an Infinite, All-encompassing Entity, then it is illogical that he could create something outside his being and reside separate from it. Because of this inconsistency, theologians today have increasingly realized the irrationality of this model for God and have begun to embrace the concept of panentheism. The panentheist concept of God has him as both the physical universe, or "in" his creation, and transcendent to it. In this way, panentheism differs from pantheism, which is the belief that the creation is identical to God. It could be argued that this movement of religious doctrine away from strict dualism (we and everything else are separate from God) is partly a result of the scientific discoveries and the rise of secular rationalism that have taken place in the last one hundred years.

Of course, Eastern religions have never had a problem with dualism. It was always a matter of faith that matter like everything else was contained within the Divine Being. Therefore the principal Eastern religions are monistic—God, the Transcendental Entity, is one with his creation and continuously manifests as the creation—the ultimate reality is the One (Brahma).

Yoga

Yogis have always maintained that yoga is a spiritual science. While the ultimate goal of yoga is unification, it is said that the psychophysical and psycho-spiritual practices of this discipline were developed and perfected over thousands of years through experimentation. There is undoubtedly some truth to this assertion. The yogic poses (asanas) were developed and refined over many years, as was an understanding of how each pose affected important glands and chakras. Yogis also experimented with how various foods affected their body and their ability to concentrate the mind during meditation. As a result, they developed a "meditation diet," which is lactovegetarian. Such a diet is designated as sentient (*sattvic*) and promotes better health and a calmer mind that is more conducive to meditation.

The higher yogic practices described by Patanjali are concentration (*dharana*) and meditation (*dhyana,* sadhana), which ultimately lead to unification (samadhi). Meditation is sometimes called "intuitional science," implying that it is a scientific method of developing intuition. Here intuition means supermundane knowledge—tapping the subtle, unconscious, collective layers of mind. Whether there is any science in developing these practices is debatable. However, spiritual aspirants are often told that the mantra they were given by their guru was empowered by the master when he used it to raise his own divine spiritual energy (kundalini) from the lowest to the highest chakra.

Summary

Some of the greatest scientists of the twentieth century realized that the new theories of space, time, and matter, and the experiments that validated these theories, demonstrated that reality was far richer, nonlocal, and whole than the model described by the mechanistic, clockwork universe of Newton. Numerous experiments showed that human consciousness affected how reality manifested; therefore, it had to take precedence over matter. This was a giant leap toward framing a new paradigm that placed Consciousness as the ground substance of creation. These insights into the nature of reality are even more remarkable because they arose before it was shown that all quanta are entangled (connected). Entanglement served as the final blow to the epiphenomenalism that put matter before mind and consciousness, since it clearly demonstrated that nonlocality and wholeness were fundamental to any understanding of reality. These are essential elements of spiritual philosophy, and for the first time in two hundred years the spiritual worldview gained support from science and members of the scientific community. Today we find many scientists and authors who have argued persuasively that the materialistic worldview, which describes reality as external to us, is false and needs to be replaced by a new worldview, which describes our being as intimately connected to the very fabric of creation itself.[18]

Interestingly, while the hard sciences have increasingly moved toward integrating metaphysics with science, the opposite trend has been going on in the life sciences such as psychology and neurobiology. In these disciplines it is easy to ignore the more esoteric and weird findings of the physical sciences and not look beyond behaviorism and reductionist dogma. However, this state of affairs cannot last long. Eventually even the soft sciences will have to accept the fact that ESP is real and that it is a hopeless task to try to explain consciousness as an electrochemical process. Following the lead of early visionaries, a day will come in the future when scientists from all disciplines will embrace a new epistemology based on Consciousness instead of matter. Then society as a whole will begin to realize that Consciousness is the true base from which creation springs.

14
Karma, Happiness, and Suffering

THE WORD "MORALITY" COMES from the Latin *moralitas* and means "manner, character, proper behavior." It is a system that differentiates behaviors that are proper from those that are improper. Moral behavior by members of a civilization is important if that civilization is to endure. With the exception of a small minority of people who suffer from mental and psychopathic illnesses, humans seem to have an innate understanding of right and wrong. This intuitive ability to distinguish right from wrong is often called conscience. However, for many people having a reason to behave morally is important. Spiritual ideology offers a knowledge-based rationale for behaving ethically as opposed to a fear-based motivation.

The Law of Karma is a central concept of the spiritual worldview. It is a punishment/reward system that teaches people how to behave so that they move toward perfection. As previously stated, the simplest explanation of this law is the well-known saying, "you reap what you sow." It is the reason one should follow the Golden Rule, since good actions beget good reactions and bad actions beget bad reactions. People who understand how actions, either good or bad, inevitably return to affect them in this life or in a future life have a reason to act morally. That is, they are aware that there is no escaping the consequences of their actions. Moral behavior is in one's own self-interest.

The importance of moral behavior for progressing on the path of self-realization cannot be underestimated. In the eight-limbed (*ashtanga*) path of yoga described by Patanjali in his Yoga Sutras, the first two steps are *yama* (do's) and *niyama* (don'ts). Without morality, progress on the spiritual path is impossible since the mind will be dragged down by the weight of immoral actions. Morality is therefore necessary on the path of knowledge and bliss, and most of the world's religions put great emphasis on thinking and acting morally.

In the past people were more inclined to avoid immoral behavior because of the fear of eternal damnation. This church doctrine certainly helped keep people "in line" and discouraged them from actions that violated the Ten Commandments. However, today few people believe that they will be sentenced to eternal damnation if they sin. In addition, few believe in the existence of hell. Fear as a basis for acting morally applies mainly to a fear of going to prison.

Happiness and suffering

According to the spiritual worldview, we are God but do not realize it. Our life's purpose is to realize who we actually are. Therefore actions that reinforce our connection with the Cosmic Entity reward us with happiness while actions that distance us are punished. Actions that draw us closer to self-realization (e.g. spiritual practices and meditation) produce the greatest happiness while actions that take us further away produce reactive momenta that cause suffering. The Cosmic Entity has unconditional love for all his created beings, but the only way they can learn right action from wrong action and move forward on the path to perfection is if they are punished when they deny their divinity and rewarded when they affirm it. In other words, individuals who work to discover and develop their innate spiritual nature (dharma) experience happiness and ultimate bliss. Those who take the path toward darkness and work against their true nature do not serve the purpose for which they were created and bring unhappiness and suffering upon themselves.

Some people believe that the karma gathered due to evil deeds can be compensated or neutralized by good actions. However, spiritual ideology suggests this cannot happen. All actions, whether good or evil, cause a deformity in the unit mind. In the process of the mind regaining a flawless form, the deformity is removed by an equal and opposite mental reaction. Hence the reactive momenta caused by an evil action cannot be removed by a good action. Every reactive momentum is independent of all others and one has to experience the consequences of good and bad actions separately. It is as though one had made deposits in an account, which they must eventually withdraw. Thus the results of good actions cannot help one evade the suffering caused by bad actions. At best, the mode of experiencing the reaction can be changed and the intensity of suffering can be reduced by slowing the speed at which reactions are experienced.

It is obvious that all individuals carry a karmic burden. It could be called their "karmic bag." They take this bag from one lifetime to the next. The weight of this bag depends on how they acted in the past. The fastest way to empty the bag would be to go ahead and experience the consequences in their bag and not put any more into it. It is not enough just to perform good actions. This is why Buddha called good reactions chains of gold. Whether the chains are gold or iron (bad reactions) both bind one to this world. In the next chapter, we will discuss methods that can clear out one's bag of reactive momenta quickly without creating new ones.

Western church doctrine maintains that the soul comes into existence at the time of conception, and it is an accident of nature if an innocent child is born blind or without arms while another is born a prince. Because man is tainted by original sin, the fate of a child is left to chance because its ancestors committed a crime against God.

Spiritual ideology rejects this illogical proposition and has a simple and logical explanation for why people experience happiness and suffering. Since reactive momenta remain in the unit mind after death, the bodiless mind, in order to find expression for its remaining karmic burden, takes on a new physical body. The Cosmic Mind will find a suitable embryo to serve as the new home for the bodiless mind so that it may continue on the path to perfection. Hence the burning of old reactive momenta and the production of new ones may go on for many lifetimes before the individual advances onto the accelerated path of *vidya* and completes the journey to perfection.

Because units carry their karma from one body to the next, we sometimes witness the phenomenon in which a child is born into a life of hardship and suffering due to no obvious fault of its own. Another child may be born with the proverbial "silver spoon" in its mouth. Since we have no knowledge of a child's past lives, we remain mystified by such so-called accidents of birth. However, everything in the universe is connected, and in reality there are no accidents—only incidents. Hence, some individuals come into this world with a seeming mountain of problems to overcome while others seem destined for happiness and success. Suffering is never accidental. It is brought upon oneself in this life or from behavior in past lives. The fact that people do not remember their negative deeds from the past does not free them from having to reap their reactions. Humans are blessed that once the negative reactions are burned they are gone forever, and at a subtle mental level they are reminded not to repeat the action that ultimately brought on the pain.

Good and evil

Western religions have struggled with questions about good, evil, and suffering. The traditional model for why there is suffering originated in the Old Testament. Evil arose in the world when Satan, a fallen angel, spoke through a serpent and seduced Eve into disobeying God's command. The goal of the devil was to lead people away from the love of God and into ignorance and evil. Suffering exists because man fell from the grace of God (original sin). The punishment for disobeying God's will and following the path of evil is eternal damnation in hell. The question that is never really answered is how an all-loving, omniscient, omnipotent, omnipresent God would allow suffering and evil to exist in his creation. Today few people believe this story and it is undoubtedly one reason that church attendance is declining worldwide.

On the other hand, spiritual ideology claims that good action is an action performed with good intention and knowledge—an action that brings one closer to the goal of becoming one with God. Evil is just the opposite and is performed out of ignorance. In the simplest sense, good is the movement from crude to subtle, or the process of identifying with the Cosmic Self as opposed to the ego. We have also used the Sanskrit term *vidya* to describe this movement. Evil action may be termed *avidya*, or the movement toward ignorance or crudity. Such actions create negative reactive momenta, which lead inexorably to pain and suffering.

The evolution of unit consciousness occurs when the reflection of Cosmic Consciousness becomes clearer and greater in intensity. In this process the mind becomes subtler and more expanded. This movement toward subtlety is accelerated by good actions such as selfless service and by focusing the mind on the Cosmic Entity as opposed to crude material objects. Because of the deep connection of individual human consciousness with the limitless "I am" of Cosmic Mind, there is an inexorable tendency for the human psyche to be drawn toward Cosmic Consciousness, and every unit will eventually follow the path of knowledge and attain unification with the Cosmic Entity.

On the other hand, the more the mind is absorbed in crude objects the more the unit is dragged backward, because the reflection of Consciousness becomes dimmer with greater expression of bondage. When the mind is absorbed in pettiness and crudeness, it remains more strongly under the influence of the Qualifying Principle; as a result, the onward march of the individual is retarded. Actions that take a person against the natural flow of the creation cycle slow the evolutionary march toward subtlety and result in punishments designed to discourage such behavior in the future. Such actions performed out of ignorance can be termed evil. On the other hand, actions that reinforce one's connection with God may be termed good.

There is a continual conflict going on everywhere between good and evil, light and darkness, virtue and vice. Human society progresses through this conflict as it slowly moves from imperfection toward perfection. The struggle between good and evil went on in the past, is going on in the present, and will go on in the future.

For the vast majority of people, progress in the social sphere has produced increasing awareness of the subtle connection they have with all of humanity. Artificial barriers such as nationality, race, religion, socioeconomic status, gender, etc. slowly begin to dissolve as people adopt a more universal or humanistic model of humankind. At the same time, moral attitudes change and grow. People develop a more family-like sentiment toward other members of society and increasingly treat others with the same love and respect that they might treat a brother or sister. Finally, at a conscious or an unconscious level, they become aware that the actions they perform will inevitably bounce back to affect them. Adoption of the spiritual worldview promotes and accelerates this process of moral growth and understanding and must be encouraged if humankind is not to sink backward into the darkness of the past. There is no mistaking the fact that such forces of ignorance are at work in the world. There exist dangerous ideologies that are both divisive and destructive to human values. For an example, one need look no further than the rise of radical Islamic jihadism in the last few decades. To call the followers of these extremist ideologies Islamic is actually incorrect, since Islam is a faith based on peace and love and views the killing of innocent people as an abomination. Moreover, the vast majority of people that have died at the hands of these terrorists have been Muslim. This is an ideology based on ignorance and thus it could be called evil. It attempts to reverse the progress made in the social sphere during the last one thousand years.

For a human, the greatest evil is to deny one's own humanity. That is, to deny one's connection to God and thus one's connection to everything and everyone in the universe. Such a mindset is no different from that of an animal, which is incapable of contemplating a higher order. While animals act primarily under the influence of instinct or Cosmic Mind, man is capable of doing evil deeds that accelerate his movement toward crudity. As a result, some truly evil individuals could be considered to have sunk to a state lower than that of an animal. However, even a person whose actions are evil is inexorably drawn toward the limitlessness of the Cosmic Entity. After reaping their negative reactive momenta and undergoing much suffering, they eventually will leave the path of *avidya* and move forward on the path of *vidya* to eventual unity.

15
The Quest for Limitlessness

IF YOU WERE TO ask yourself what motivates you to live the way you do, you would probably say that you are on a quest for happiness. At the core of the human condition is the desire to be happy. If God were to grant us one wish, we would probably ask to experience unlimited and unending happiness. This is natural, since humans possess a keen sense of self-awareness and knowingly or unknowingly crave the limitless ecstasy that results from merger with the Cosmic Entity. In a human being, the "I feeling," *mahat*, is the dominant characteristic of their mind. Since *mahat* is nonlocal and unlimited, the personal "I feeling" is identical to that experienced by every other human being, and it is through this infinite "I feeling" that everyone shares an intimate connection to the Cosmic Entity. This connection leads to an underlying longing for the great and to a thirst to experience unlimited happiness.

The longing for limitlessness is a human characteristic not observed in animals. Animals can be completely satisfied by a full meal and then fall asleep. Humans never seem to be completely satisfied with anything on the physical plane because nothing physical can completely satisfy their craving for limitlessness.

The underlying thirst for the Infinite can literally drive people to act in destructive and dangerous ways, especially when this energy is directed toward the material realm. For example, the sickness of accumulation and greed can occur when a person devotes their life to the acquisition of wealth. After acquiring a million dollars, they remain unsatisfied and unhappy and feel the need to acquire ten million. Once that goal is reached, they still feel unsatisfied and inferior to others who are wealthier. Hence, they devote their energies to acquiring hundreds of millions of dollars or perhaps billions of dollars. After a person has obtained fantastic wealth, they may still feel unhappy and unfulfilled. It seems that nothing less than the acquisition of the entire universe would bring them complete satisfaction. Sometimes people develop the desire to climb the highest mountain, be the greatest athlete, or excel at some other physical endeavor. If they are not injured or killed, such devotion of time and energy to external activities may ultimately lead to feelings of accomplishment and happiness once the goal is achieved. However, those feelings are short-lived and never completely satisfying.

The thirst for limitlessness may also be misdirected into the quest for fame or power. The inflated egoism that accompanies these endeavors brings only fleeting happiness; such people never achieve complete satisfaction because no matter how much fame or power they attain it is less than absolute and always short-lived.

Other people try to become intellectual giants or foremost experts in their field. This is not destructive unless it leads to intellectual pride or self-righteousness. Still, their energies are focused on the lower plane of existence when they could be directed to the limitlessness of Spirit.

It is obvious that happiness is a state of mind that comes from within. Confusion arises when one associates happiness with outside activities and objects. The activity or object itself contains no happiness—they merely stimulate those feelings in the mind. When one ties all their hopes and dreams for happiness to sensory pleasures and material objects, which bring only temporary happiness and are never completely fulfilling, disappointment is inevitable, leading to increasing frustration in the search for that something special. Not unlike an addict that needs increasing amounts of a drug to provide the same high, the search for happiness "out there" leads inevitably to unhappiness and dissatisfaction with life.

On the other hand, there is no limit to the riches of mind and Consciousness. These riches are obtained not through the sense or motor organs but through introspection—i.e. self-knowledge. Happiness as opposed to simple pleasure comes when the mind is engaged in the introspective exploration of the limitlessness of the psycho-spiritual realms, and true happiness is obtained when the mind is fully engaged in the limitlessness of Consciousness or pure being.

Human thirst for the infinite is insatiable and every human being craves happiness—not a small amount of happiness, but infinite happiness; and not in the distant future but right now. The only thing that can quench this thirst is the realization that they are God. To attain such self-realization one needs to first enter a path of self-discovery, and the most effective practice for this purpose is meditation.

The power of meditation

When people finally realize that they are already on the path to perfection and desire to be guided by knowledge rather than by happenstance, they inevitably begin in earnest to perform some type of spiritual practice. It is with such practice that one can move at an accelerated rate on the *vidya* path of bliss.

Many practices have evolved over the ages for developing an awakened state of consciousness. Examples include religious rituals, prayer, practicing humility and service, love and reverence for God, chanting, and self-surrender. Most are designed to draw one nearer to God and to focus the mind on him instead of the outside world. However, to become fully absorbed in the Cosmic Entity the mind must become one-pointed or concentrated. Of all the various spiritual practices, meditation is probably the most powerful technique for achieving self-realization. The goal of meditation is nothing other than cosmic ideation, floating on the divine waves of bliss, drawing ever closer to the Cosmic Entity. However, meditation practices are subtle in nature and can be difficult. They may not give immediate results and are best learned from an experienced teacher.

Meditation practices probably had their origin over ten thousand years ago in what is now southern India. The practices were greatly refined and improved by the great Tantric sage Sadashiva about seven thousand years ago and became incorporated into the various mystical traditions of the East, including yoga, Tantra, Buddhism, Hinduism, Taoism, and Sufism. Spiritual practices were probably first introduced to the West as early as ancient Egypt, whose religion appears to have been influenced by Asian mysticism. However, it was not until the beginning of the twentieth century when Swami Vivekananda visited the United States that meditation practices became better known in the US. He was followed by numerous other teachers of Eastern mysticism, especially during the 1960s when there was an explosion of interest in all things Eastern. This "new age" movement was partly fueled by the use of psychedelic drugs (e.g. LSD), which offered to throw open the "doors of perception" and for a short time produce experiences similar to a mystical experience.

The Yoga Sutras of Patanjali succinctly describe the steps needed to achieve the goal of union (samadhi). First moral behavior is required because without this the mind cannot practice introspection. The physical exercises (asanas) help tone both the body and its glandular systems. These practices also help calm the mind and prepare the body for sitting comfortably in meditation. In addition, the yogic diet of fruits, vegetables, grains, beans, nuts, and milk products helps prepare the body for the higher sentiments produced by meditation. Next, breath control (*pranayama*) helps control and balance the *prana* or psychic energy. Slow rhythmic breathing calms the mind and prepares it for meditation. Even simply sitting comfortably and watching the breath calms the mind and is perhaps the simplest form of meditation. Patanjali broke meditation down into three phases. The first (*pratyahara*) involves withdrawing the mind from the activities of the sense and motor organs, which allows it to focus internally (*dharana*), after which the mind can attempt to associate with the Cosmic Entity (*dhyana*).

In the broadest sense, meditation is any practice in which the ego is set aside, causing one's sense of doership to disappear as the mind becomes detached from bodily sensations. In the process, the person becomes lost in what they are doing, whether it is playing music or becoming fully engrossed in trying to solve a problem. When practiced for the purpose of spiritual growth the only difference is that the object of concentration is the Cosmic Entity. However, the Cosmic Entity is purely subjective and for this reason is sometimes called the "Supreme Subjectivity." This creates a difficult problem for anyone trying to practice meditation for the purpose of self-realization. The mind has a natural tendency to want to focus on something objective, but if the goal of spiritual practice is to transcend or extinguish the mind, then it has to become absorbed in "Supreme Subjectivity." Because of this conundrum, spiritual teachers have developed different techniques that aid a spiritual aspirant in their quest for limitlessness and divine bliss.

In the distant past, teachers of intuitional science began by determining that prospective disciples were well established in moral behavior. Then they might be taught simple concentration exercises, such as concentrating on a candle flame or a religious idol. Because those teachers were aware that meditation practices could be used to do great harm as well as great good, it was often years before the disciple was given more advanced meditation techniques.

Today most people are well established in morality and are ready to profit from practicing meditation. A simple technique that anyone can do is mindfulness meditation. In this practice, the focus of the mind shifts from attention at its surface, which is in a constant dance of change, to a deeper experience of silence and calmness. Mindfulness means being aware of the observer in us. To perform simple mindfulness meditation, one sits comfortably in a place free of distractions, closes one's eyes, and lets the mind focus on the breath. Trying to remain in the here and now, any thought, image, or sensation is allowed to freely come and go without trying to push it out of the mind or paying any attention to it. If distractions come then the mind is brought back to its focus on the breath.

In a recent study, even this simple meditation practice was shown to make significant changes in the brain. Anatomical magnetic resonance images were obtained from sixteen healthy, meditation-naïve participants before and after they underwent an eight-week mindfulness meditation program. Changes in gray matter density were measured and compared with a control group of seventeen individuals. The study found that an average of twenty-seven minutes of daily practice of mindfulness meditation produced a significant boost in gray matter density, specifically in the hippocampus, the area of the brain in which self-awareness, compassion, and introspection are associated. Furthermore, this boost of gray matter in the hippocampus was directly correlated to a decreased gray matter density in the amygdala—an area of the brain known to be instrumental in regulating anxiety and stress responses. In contrast, the control group did not experience changes in either region of the brain, thus ruling out the possibility that the changes observed were due to the passage of time.[1]

Another effective method of meditation involves using a mantra. *Mantra* is a Sanskrit word that means "the quality that liberates the mind." Hence mantras are typically Sanskrit words or phrases that have particularly strong spiritual vibrations and a spiritual meaning (often a name of God). The use of a mantra for meditation takes advantage of two important qualities of mind. First, that the mind requires an object—simply trying to empty the mind of any thought or impulse is very difficult, if not impossible. A mantra serves as the object of concentration, and since it has a subtle vibration and since the mind takes on the qualities of the object of its attention, the mind becomes subtle. Second, since the mind can only think of one thing at a time, if it can focus on the mantra then other thoughts and sensations cannot enter. Normally people experience a lot of activity in their mind. Such mental activity is actually unilateral (one thought at a time), much like the fast-moving frames of a motion picture. It just seems like the mind is thinking and experiencing multiple things simultaneously, but similar to a motion picture, what appears continuous is in fact made up of individual snapshots with little gaps in between. The great sage, Ramakrishna, once likened the mind to a mad monkey that continuously jumps around. Then he corrected himself and said, "no, more like a mad monkey bitten by a scorpion."

A mantra may be repeated silently (*japa*) or sung out loud (kirtan). The mind is naturally attracted to the sweet vibration of the mantra, and ideally, it is drawn away from the chaos of the external world into a deep state of inner concentration and peace. A mantra may be considered a spiritual tool that is able to bridge the void between the objective level of existence (mind) and subjective reality (Consciousness).

The two most common types of mantra used for meditation are a seed mantra (*biija*) and a "God" mantra (*ishta*). A seed mantra may be one of the sounds associated with a chakra. By repeating the mantra silently, the mind is drawn to a deep level below where thoughts originate. The mantra may also be designed to open up a chakra, such as the heart center, allowing love to flow through one's being.

The God or *ishta* mantra is central to the practice of Tantric meditation. It is a two-syllable Sanskrit word meaning "God." The first syllable is repeated silently while breathing in and the second is repeated while breathing out. It is necessary to go to a specially trained teacher of meditation to receive an *ishta* mantra since it is tailored for each person based on his or her mental and spiritual propensities. An *ishta* mantra is instilled with strong spiritual vibrations by a spiritual teacher or guru. This mantra is especially designed to cut through the outer layers of the mind and stimulate the dormant spiritual energy (kundalini) that lies in the lowest chakra. This energy then rises up the principal channel of the psychic body, purifying each of the chakras above it, and ultimately brings spiritual energy and bliss to the practitioner.

In the final analysis, all forms of spiritual meditation have a single purpose—to direct the mind away from the crude world of the sense and motor organs toward the Cosmic Entity. In the process, the unit "I am" and "I do," which are associated with ego, become associated with the Cosmic "I am," which is an egoless state. The resulting mental expansion is accompanied by a new understanding of the meaning and purpose of existence and is accompanied by bliss.

Although meditation is the key practice for accelerating one's movement on the *vidya* path, everyone can profit from practicing it. In the absence of spiritual practice, individuals move slowly on the path toward perfection—learning from past mistakes, expanding their mind slowly by the constant struggle and resulting growth that comes from life experiences. In contrast, by practicing meditation individuals experience both inner peace and more rapid changes in their life as their mind expands and they move closer to self-realization. They also experience numerous other physical and psychological benefits. For example, aside from increasing brain gray matter, meditation has been shown to do the following in scientific studies:

- Lower blood pressure.
- Lower the levels of blood lactate, reducing anxiety attacks.
- Decrease tension-related pain, such as tension headaches, ulcers, insomnia, muscle, asthma, and joint problems.
- Increase serotonin production, improving mood and behavior.
- Improve the immune system.
- Increase energy, as one gains an inner source of energy.
- Bring the brainwave pattern into an alpha state that promotes peace, healing, and pain relief.

Other reported physical benefits include:

- Keeping one stress free.
- Slowing aging.
- Adding more hours to the day by reducing the need for sleep while improving the quality of sleep.
- Helping one appreciate life more.
- Helping one feel more connected to other people and living things.
- Making one happier and those around them happier.
- Improving the functioning of the brain.
- Improving metabolism and helping one lose weight.

Mental benefits of regular meditation include:

- The mind becoming fresh, delicate, and beautiful.
- Decreased anxiety.
- Improved emotional stability.
- Increased creativity.
- Increased happiness.
- Developing intuition.
- Gaining clarity and peace of mind.
- Increased self-confidence, self-awareness, and optimism.
- More harmonious relationships with friends, family and colleagues.
- Problems become smaller.
- Sharper mind with improved focus.
- Improved emotional steadiness and harmony.

The spiritual benefits of meditation may be summarized as:

- Bringing about a true personal transformation and knowledge of who one really is.
- Gaining a deeper sense of purpose and a happier, more fulfilling life.
- Attaining indescribable bliss.
- Liberation from the endless cycle of birth and death.
- Achieving self-realization or enlightenment.

16
The Problem with Ego

Throughout human evolution, threats to our survival have been met by an activation of the autonomic nervous system, which releases adrenaline and prepares us to fight or flee from danger. The self-preservation instinct that we share with animals was necessary for our survival as a species, but it is no longer as important for modern man. Nonetheless, we still tend to react to minor threats to our ego as if we were in danger of bodily harm. For example, someone criticizes us, hurts our feelings, or does not give us the respect we think we deserve; we instantly want to turn away from that person or shoot back with some biting remark of our own. The pain to our ego may last a long time and we may shun further contact with that person and malign them to others.

It is the nature of ego to create separation between self and nonself, since it functions in the realm of "me" and "mine." Hence one of its jobs is to create separation. The ego provides the sense of willfulness ("doer I" or *aham*) that exists behind all that a person does, says, thinks, understands, or hears. It is at the root of selfishness, judgment, rejection, and separateness. Because it wants to control, it also wants to dominate the nonself, which includes other humans, animals, and nature (matter and energy). The ego is threatened by the idea of losing control. Hence death is the greatest fear of the ego, since death brings loss of control and loss of individual consciousness. Since these occur also in self-realization, the ego is deathly afraid of merging with the Cosmic. It will seemingly do everything in its power to prevent what it perceives as a loss of control. Hence ego can be a great enemy for spiritual progress. It has been likened to an umbrella blocking Spirit and preventing the mind from experiencing God's grace, which is continually raining down on us.

However, ego is an indispensable component of a healthy mind, for without the "I do" engine of ego, one would be like a vegetable, unable to make any movements or decisions. Therefore the goal of the spiritual aspirant is not to destroy the personal ego but to wear it down and rely more and more upon cosmic ego to perform actions. Personal ego is the main barrier for perceiving the oneness that lies beyond. Hence personal ego has complementary aspects: on the one hand it allows one to function in the world, and on the other it perpetuates the illusion of duality and individuality. A solution to the problem of ego might be to renounce this world as illusory and retire to a cave in the mountains to practice meditation. Most spiritual paths today reject such escapism. They teach that the relative world of duality is a reality that must be dealt with and can be utilized for the purpose of transcending personal ego and drawing nearer to God.

Personal ego is boosted by achievements, such as wealth, fame, and power, and the more achievements attached to the ego the further one is led from the pure essence of the "I am" state of mind. It is the source of people's separation from God, the source of their torments, pain, desires, anxiety, frustration, numbness, attachments, and isolation. When active, there is no escape from the constant mental disturbance that results from knocking the ego against those of other people and against the universe in general. The good news is that such struggle can cause the ego to become subtler as it is converted into "I feeling," resulting in mental expansion.

Ego development

A baby is born innocent. It has no ego, no individuality, and no discrimination between self and nonself. The baby is oblivious to problems and attachments, totally accepting and trusting of its world. It knows only love. It lives totally in the timelessness of the present moment. This state of pure, unconditional being is similar to the egoless state of consciousness that a seeker strives to attain in life. However, for a baby this perfect state of bliss and contentment does not last. Moreover, a child is not born as a blank slate. Each child brings its unique karma—traits that give the child its character and individuality. The mother instantly recognizes her baby from all others and knows its peculiarities. Even very early on in a child's development, remnants from their previous life begin to color their individuality and self-image. However, their consciousness of self is undeveloped along with their memory, which for a young child is noncontinuous or sporadic. Only a few memories are retained prior to the third year, to say nothing of a previous life, but reactive momenta differ from memories in that they originate from a deep unconscious layer of mind and directly affect the physical body and mind.

Gradually, as needs and desires arise that are not immediately fulfilled, pain arises and the beginnings of fear, anger, desire, attachment, and doubt emerge. In order to cope with the physical world the child gradually learns the difference between self and nonself and begins to speak of itself in the first person. The developing ego imparts a sense of duality—the difference between me and it or me and you. The ego separates what is physically real from what is unreal. It helps the child organize its thoughts and make sense of them and the external world. This development is essential for the healthy growth of the child, for without it the discrimination and judgment needed for survival would be lacking.

As the child develops mentally and physically it develops empathy, learns social rules of behavior, and forms friendships. However, the child's character is not solely a product of their nurturing. A good portion is inherited. Inexplicably the child may take a life path that is unexpected and very different from that of its parents. This is because the individuality of a child is dependent on both the parental family and environment as well as upon the karma it carries from previous lives, which need expression in this life.

Ego development continues into and beyond adolescence as strong emotions of sexuality, love, dislike, and anger take hold. Ego development does not end here, however. Most people today believe that happiness is a result of their accomplishments. To be happy one needs a meaningful and well-paying job, a spouse, children, a nice home, and personal conquests. The ego's quest for achievement has no real boundaries because underneath the "I do" mentality of ego lies the limitlessness of the "I am" realm. However, as the ego amasses layer upon layer of possessions, wealth, status, and power in its search for happiness, its burden grows. Possessions are lost or lose their attractiveness over time. A lack of fulfillment and a sense of dissatisfaction with life may develop.

As old age creeps up and the motor and sense organs become less adept and keen, external activities do not bring as much pleasure as in youth. Trapped in the domain of me and mine, people dominated by ego fail to experience the true happiness that lies beyond the material realm. To graduate from ego attainment to seeking union with the Infinite is a normal and healthy change that takes place as a person ages. However, for many people the transition never takes place. They remain attached to the pleasures of the body and physical world. They fail to learn of the limitless riches gained by transcending ego and searching within for peace and happiness as well as the answers to life's mysteries. In his book *Modern Man in Search of a Soul*,[1] Carl Jung described the problems such people encounter as they age. They try to hold onto the pleasures they experienced as a youth as they enter the second half of life. Naturally, they are accustomed to acquiring such pleasures in the external world, but their sense organs inevitably become duller and the motor organs weaker. Such people seek greater and greater stimuli in order to compensate and may fall prey to what is termed a mid-life crisis. If left uncorrected, this attempt to stem the tide of old age can lead to an autumn of life marked by discontent, dissatisfaction, cynicism, and unhappiness. No matter how hard one tries, it is impossible to reverse the physical effects of aging. Jung argued that to be mentally healthy as a person ages they need to give serious attention to their psycho-spiritual development. Being freed from many of the mundane obligations of youth in their middle years, they have the opportunity to reap the incredible treasures that an introspective approach to life can provide—for true happiness springs from within, not from without.

The ego is only active in the conscious and subconscious layers of mind. It is constantly involved with the mundane tasks of everyday life. There is no reason for the higher level of mind or "I feeling" to compete with the ego since this level of mind is already fully engaged in Cosmic Consciousness. A positive effort must be made to transcend ego. This means attempting to direct the mind toward the Cosmic Entity. Inevitably, this creates struggle and mental conflict because the ego wants to maintain control and sees the loss of individuality that accompanies self-realization as equivalent to death.

The mutative binding force, which creates the sense of "I do" in Cosmic Mind (Cosmic Ego), is reflected on the mind of individuals as personal ego. While it is essential for the first phase of human development, the vast majority of people remain trapped by the selfish allure of ego and do not graduate to the next level—that of the seeker.

The seeker

Because humans are connected with Spirit, they experience an inner voice that speaks to them in subtle whispers. This voice is only quieted by the constant activity of the conscious mind or the stupor of unconsciousness, such as that brought on by drugs. When an individual finally comes to the realization that there is more to life than the incessant quest for pleasurable experiences and achievements, then they inevitably start down the path of a seeker. A seeker is thus someone who has entered the *vidya* path of self-discovery. A seeker has acquired wisdom and realized that knowledge of self is knowledge of God. Such a person rejects the idea that extroversive activities bring lasting happiness. They have come to realize that true happiness comes from within. A seeker has begun the search for the meaning of life and the attainment of self-realization. The major concerns of the ego are put aside as the seeker begins to give themselves to other people and develop feelings of universal love for humankind. The seeker has developed an understanding that there lies a far deeper and richer reality beyond that of the ego and the experiences of the sense and motor organs. In a sense, a seeker has crossed the threshold from the stagnant waters of a pond into the flow of a stream that will lead eventually to the ocean. Seekers are characterized not only by an interest in spirituality but also by the practice of some form of introspection such as meditation. In their search for truth, such individuals become tolerant, nonjudgmental, loving, selfless, service minded, more in tune with their body and nature, healthier, and happier than they ever thought possible.

Not every step on the seeker's path is blissful since they are still encumbered with negative karma from the past and plenty of ego baggage to overcome, but the rewards they feel along the way offer convincing proof that they are on the right path. According to Deepak Chopra a seeker is giving by nature and wants nothing in return, not even gratitude, being motivated by selfless love and compassion. Intuition becomes a trustworthy guide, replacing strict rationality; they catch glimpses of an unseen world as the higher reality; and intimations of God and immortality appear. These signs are accompanied by growing enjoyment of solitude, by self-reliance in place of social approval, by stirrings of being and a willingness to trust.[2]

Various terms are used to describe a seeker. These include *sadhaka*, spiritual aspirant, chela, adept, yogi, Tantric, monk, nun, acharya, disciple, devotee, and spiritualist. A seeker may endeavor to overcome ego but is still largely under its influence.

The seer

The most advanced or subtlest phase of life is that of the seer or sage. The seer sees everything as a manifestation of God. For such an individual the illusion of separateness breaks down. For the seer, unity is not something taken on faith or a belief—it is experienced. The seer has surrendered their ego and experiences the world in its pure form—as manifest Consciousness. Hence the seer feels at one with every living creature and with every particle composing the universe. The seer is completely open, no longer plays psychological games, and is incapable of feeling any emotion except unconditional love for God and his creation. The seer lives totally in the now and has access to the unlimited knowledge that lies in the collective unconscious or Cosmic Mind. Seers are not attached to the fruit of their actions; they feel that they are merely instruments for the will of the Cosmic Entity. Thus a seer is free, living in a state of grace and indescribable bliss.

It is natural for people to be drawn to such individuals and look up to them for advice and spiritual guidance. The seer is sometimes called a saint, spiritual teacher, guru, rishi, sadhu, master, sage, or prophet. The sentient quality of the Qualifying Principle ("I am" or *mahat*) dominates this final phase of human development. Knowingly or unknowingly all human beings crave the limitlessness that accompanies this, the ultimate phase of life.

Going beyond ego

God is the ultimate Good, and if one could experience the egoless state, which is characteristic of the highest level of mind, they would feel connected to him and bask in the ecstasy of his being. In reality, humans are no more separate from the Source than rays of sunlight are separate from the sun, but ego gets in the way by creating the illusion of separateness. Ego obscures the Ultimate Reality (Consciousness), covering it with layer upon layer of "I am so and so," and "I do such and such." Ego is the source of arrogance and suffering, and to believe that expanding its power should be the goal of life is the epitome of ignorance. However, getting rid of it is very difficult. It is a close and entrenched part of the mind. The question is what practices can one employ to overcome or transcend ego?

One method, which could be called the "hard way," is simply to go on living life on the surface of one's being. Inevitably, one will learn from the struggles of life experiences that it is necessary to subjugate the ego to attain happiness. Often it is feelings of unhappiness, dissatisfaction, pain, and suffering that cause people to change the direction of their lives. Hence, suffering can be a blessing in disguise, and great suffering can beget great growth. This process entails the burning of negative reactive momenta, which brings mental anguish, pain, and suffering, but in the process the ego becomes subtler, yielding more of the sentient "I feeling." For some people these experiences are what are needed to turn them away from the false promise of ego attainment to the path of true happiness. For the rest, growth will be slow but steady, possibly taking many lifetimes.

The easy way to reduce personal ego is to identify with the witness, instead of the performer. In other words, to make an effort to direct the mind toward the Cosmic Entity—the witness of creation. This is the purpose of spiritual meditation, and this effort is rewarded with happiness. If pain and suffering arise, as they inevitably will, they have less effect on the mind because the ego is strongly attached to the body and mind, and weakening the ego weakens this connection and causes a sense of detachment from pain and suffering. At the same time, meditation promotes feelings of freedom and happiness.

In addition to the practice of meditation, surrendering the ego can be an effective method of freeing oneself from its grip. This is taught by most of the world's great religions in order to reduce people's sense of separation and to increase their love of God. For example, Jesus taught unconditional, self-sacrificing love for God and for humanity. He preached service, humility, and forgiveness. He is quoted as saying that one should turn the other cheek if someone slaps you;[3] to love your enemies and pray for those that pursue, slander, and falsely challenge you.[4] He said that one must become like a child to enter the kingdom of God, completely loving and trusting.[5] It is obvious that Jesus knew that the greatest barrier separating an individual from the Lord is their personal ego. His teachings, if followed with real awareness, greatly diminish the ego's control over one's life and allow one to see through the illusion of separateness from God.

It is easy to talk or read about surrendering the ego and becoming detached from its worldly demands. It is another thing to do something about it so that one can achieve the unity they seek. The ego is tenacious. It disguises itself in desire and attachment. It will never give up control without a fight, and attempts to crush the ego by acts of self-flagellation, self-denial, self-denigration, or surrender to the will of other persons are counterproductive. Every act of surrender and service should be to the Cosmic Entity, and service should be performed without the thought of gaining recognition or thanks. Such selfless service to the manifestations of the Cosmic Entity serves to diminish the personal ego, and as a byproduct, service brings happiness to those who are served and to those who serve.

However, there is one knotty problem with performing actions, including service to humanity. Actions performed using the ego, which means virtually all actions, create new reactive momenta. Even new momenta created by good, service-minded actions will have to be burned before the mind can obtain perfect peace. So what can be done? The answer lies in knowing that actions performed without personal ego but utilizing cosmic ego do not create reactive momenta. In other words, if an action is performed with the ideation that the Cosmic Entity is performing the action, then the mind suffers no reaction to that action. The thought can be, "I am the machine, and he is the machine operator." Alternatively, one can think that they are nothing but a wavelet in the mind of God, that God is performing the action. The Sanskrit term for this type of ideation is *madhuvidya*, which simply means ascribing godhood to every living organism and object. Hence *madhuvidya* is a form of ideation upon the Cosmic Entity. It is a practice that allows one to carry on a normal worldly life and not create additional karma. A few practical examples might help. When gardening, one might try to feel that they are aiding in the growth of manifestations of God in the form of plants. Alternatively, when walking down the street one might try to see other people not as strangers but as brothers and sisters—manifestations of the Cosmic Entity. By trying to see every living organism and inanimate object as a manifestation of God and serving them as though one were serving God, one's exterior and interior will be filled with cosmic bliss and all afflictions will be extinguished.

Summary

The Cosmic Entity sees everything—everything is within his mind. He is the Supreme Subject and everything else in creation is his object. In spiritual meditation, the mind is directed toward this Supreme Subjectivity. Meditation helps pierce the veil of ego and weaken its hold on the mind. This practice creates feelings of peace, harmony, and nonattachment to the results of one's actions. The other practice that helps transform personal ego into cosmic ego and greater spiritual realization is selfless service—service with the ideation that one is serving God. Unlike meditation, which is solitary, such service helps others obtain social, economic, and spiritual growth. Hence the motto of a seeker can be "realization of self and service to others."

Western scientific reductionism and religions do little to invalidate the myths of separateness and materialism. As a society, we pay a high price for subscribing to these dogmas. The majority of ordinary people live by the principle of separateness. Living only on the lower planes of existence, they identify themselves with their bodies and lower minds. In this state of ignorance, they see themselves as separate from the world and from other human beings. They erect social and psychic barriers between themselves and others—barriers such as nationality, race, gender, religion, and economic status.

PART IV

CAN THE SPIRITUAL WORLDVIEW SAVE HUMAN SOCIETY?

Most of the problems and conflicts that humanity faces today can be traced directly or indirectly to the adoption of the materialistic worldview and to ignorance of the spiritual worldview by scientists, intellectuals, and the public in general. These problems arise from the ideology of separateness that materialism propagates. In this final section, we will argue that science has essentially proven that nonlocality is a fundamental aspect of reality and this indicates that God as Consciousness exists. Then we will discuss the specific problems that face civilization today and how spiritual ideology can solve these problems.

We will argue that what is needed if human society is to survive is nothing less than a paradigm shift in human awareness, away from the bottom-up materialistic worldview and toward the monism of the spiritual worldview. Besides the acceptance of materialism by most scientists and intellectuals today, the spiritual worldview has two other hurdles to overcome before it can excite the imagination of the public. One is the rampant consumerism that permeates the economically privileged societies of today—i.e., the overriding concern that people have for possessions, material wealth, and physical comfort to the exclusion of intellectual and spiritual pursuits. The second hurdle is religious dogma and attitudes that are in direct opposition to spiritual ideology. The latter might present an even greater challenge to the spiritual worldview than materialism and consumerism, since religious doctrines often discourage the faithful from diving deeper than the surface of their being and create a mentality that is closed to new ideas. After all, the path was established long ago by the "prophet" and success on the path depends only on good behavior and faith.

Today the trend is toward secularism and away from the philosophy of materialism. This bodes well for the eventual adoption of the spiritual worldview, but there is still a question whether such a paradigm shift will come in time to save civilization.

17
Nonlocality and the Existence of God

TODAY THERE IS NO longer any doubt that nonlocality is a scientific fact, and any description of reality must account for nonlocality. This is why the materialistic worldview is false. It cannot account for this fundamental property of reality. Moreover, if the bottom-up ontology is false, then the alternative, top-down ontology of spiritual ideology must be true. Let us begin by reviewing the evidence for these bold statements, beginning with evidence from the physical sciences.

The nonlocality of matter and energy

- *Quanta display spatial nonlocality.* Quanta cannot be placed with certainty at any given set of coordinates. They are described mathematically by a wave function, which provides only probabilities of where they will be found when observed or measured.
- *Quanta display temporal nonlocality.* The wave function is a timeless realm. It exists outside space-time, and the past, present, and future are meaningless when discussing this realm that underlies physical reality. It is only after the wave function is collapsed by observation that an arrow for time comes into existence. Experiments such as the delayed-choice experiment where the effect precedes the cause have confirmed this.
- *Quantum nonlocality is convincingly demonstrated by experiments testing Bell's theorem.* Entangled quanta remain in intimate contact

with one another even if they are on opposite sides of the universe, and communication between them is instantaneous or occurs outside linear time.

- *Quanta demonstrate wholeness.* Since we cannot know the state of a particle before it is measured, quantum theory concludes that it must be a superposition of all possible states. Before observation, quanta must be considered as all encompassing, everywhere at the same time, since they can pop out into actuality anywhere in the universe. This web of connectivity permeates the entire universe and is a hallmark of nonlocality and wholeness, which is characteristic of the quantum realm. Therefore the wave equation describing a quantum system must be a fundamental aspect of reality that exists at a deeper level than ordinary reality and corresponds to Cosmic Mind—the mind of God. It is only after a conscious act of observation that "parts" appear in material reality out of the wholeness.
- *For a quantum system, the observer and the observed system cannot be separated.* The observer or his instruments are part of the system and influence the outcome of the observation. In other words, the act of observing alters or influences the system, and this alteration occurs outside linear time. The common-sense notion of cause and effect does not apply to quantum systems. For example, the wavelike or particle-like behavior of a quantum particle is determined by the decision of the experimenter, but the two complementary aspects of the quantum (i.e. wave and particle) cannot be simultaneously observed.
- *Prior to observation, quanta exist only as possibilities.* This means that without an act of measurement, observation, or consciousness, any atoms that consist of elementary particles remain as possibilities, as do any molecules, living cells, or brains. In other words, from a bottom-up perspective there would be nothing to collapse the wave function into actuality. Something outside the quantum mechanical system is required before the system can manifest physically.
- *The brain is a quantum mechanical system.* Nothing physical could cause the wave function of the brain to collapse, because it would be part of the system. Without such a collapse, not even a single neuron could fire. Observation of the quantum mechanical brain by mind and/or consciousness is necessary before it can leave the realm of potentia (realm of the wave function) and enter the realm of physical reality—no exchange of energy is needed.

- *Everything on the physical plane has an associated wave function.* This includes the universe as a whole. Quantum mechanics postulates that without Consciousness the universe would exist only in potentia, thereby confirming top-down or spiritual ontology.

The nonlocality of space-time

Albert Einstein was the first to propose that time and space are not separate but form a four-dimensional continuum called space-time. Numerous observations and experiments have verified his theory. The current scientific understanding of space and time is inconsistent with the bottom-up, materialistic worldview because science now describes space-time as one, an unchanging whole—consistent with the spiritual worldview.

- *Time and space are intimately interconnected.* They create a four-dimensional continuum in which space is observed to convert to time and time to space.
- *The space-time continuum describes a universe that is whole.* Beneath the ever-changing reality of human experience lies a deeper, more subtle, unchanging four-dimensional realm of reality that is whole and which is termed space-time. It is an all-encompassing, timeless domain of which our three-dimensional world is merely a shadow.
- *Things change in time but they cannot change within integrated space-time.* Hence events do not take place in time—they simply are. All events that have happened in the past or will happen in the future are already there in four-dimensional space-time. This means that our everyday experience of the flow of time from the past to the present to the future is an illusion.
- *As an object moves through space, it compresses space, and time for that object slows down.* Light cannot move through space-time any faster than the speed of light since it is limited by the compressibility of space. It is impossible for a physical object to attain the speed of light, because the mass of an object increases as it nears the speed of light, theoretically reaching infinite mass at that speed.
- *Within integrated space-time only events have meaning.* An event that took place in a distant galaxy, such as a supernova, might be witnessed on Earth twenty-five million years after it took place.

However, from a space-time perspective the two events occur simultaneously and the time difference is simply a measure of the space-time coordinates separating Earth from the distant exploding star.
- *Movement causes space to convert to time.* When an object is not moving then it is moving in time alone. If an object is moving near the speed of light then it is moving mostly through space and its time will slow down.
- *Gravity is simply a function of the bending of space-time.* A massive object distorts space-time much as a bowling ball would bend a membrane of rubber. Gravity pulls on time and it is slowed down similar to an object moving near the speed of light. Light traveling through curved space is bent. This phenomenon is called "gravitational lensing." In a massive black hole, the curvature of space is so great that light cannot escape. The things we call space, time, gravity, and light are clearly entwined.
- *Space and time are observer dependent.* Time and length may expand or shrink depending on the relative state of motion of the observer and what is observed. As space shrinks, time dilates. Space is transformed into time and time into space.
- *Space-time contains enormous energy.* Subatomic particles are observed to continuously emerge and disappear into the "vacuum" of space-time. The energy of space-time, known as dark energy, constitutes about 70 percent of the total mass-energy of the universe. The energies for normal and dark matter combined comprise the other 30 percent.

These strange properties of space-time are entirely consistent with nonlocality and spiritual ideology. We will go through the spiritual explanation for these properties one at a time.

- *The wholeness of space-time reflects the wholeness of the created universe.*
- *The timelessness of space-time reflects the reality of a domain that is subtler than ordinary reality.* This domain penetrates and surrounds ordinary reality but is not localized in any part of space-time because it is everywhere at the same time—reflecting wholeness. This subtle realm can be called Cosmic Mind, and it is witness to all four-dimensions of space-time, which are unchanging or fixed in what can be called the eternal now.
- *Spiritual ideology postulates that Consciousness is transformed into mind and then into matter-energy.* The speed limit for light (and

matter) is required to maintain the separation of mind and matter. Theoretically, if matter-energy (physical phenomena) had no speed limit, they could enter the subtle realm of mind, which is characterized by wholeness. Matter and energy could be everywhere at the same time. A speed limit prevents this from happening and prevents objects from moving backward in time, which would create impossible paradoxes.

- *Space, time, gravity, and light are entwined.* Mass and energy are similarly entwined and these connections result from the underlying wholeness that is at the foundation of reality.
- *Reality is observer dependent.* Without Consciousness, the universe could not come into actuality. Therefore it follows that space and time are observer dependent in the same way that quantum systems are dependent on observation before they can become physically manifest.
- *Space-time has a huge amount of energy locked in it.* It is the subtlest of the five fundamental factors that compose the physical universe. According to spiritual ideology, the greatest power or energy lies in the subtle as opposed to the crude. Hence space-time has more energy than all the other matter-energy of the universe combined.

Complementarity indicates underlying unity

Complementarity is an important principle for understanding and acquiring scientific knowledge about the nature of reality. An example is the dual aspects of light as both wave and particle. Depending on the experimental design, light may behave like a wave or like a particle. Neither description works under all experimental conditions. This indicates that the true reality for light must lie at a deeper level than that of the two mutually exclusive descriptions. The dual nature of light is analogous to waves on the surface of the ocean. The waves are simply the outermost expression of the ocean. A full description of the ocean would have to include the depths that reside below the surface.

Complementarity in the quantum realm indicates that there is a subtler level of reality behind the observed physical reality. In the case of light, it would be the nonlocal wave function. Other complementary constructs in the physical realm are matter-energy, space-time, position-momentum, potential-actual, and cause-effect. The concept of complementarity can also be applied to other constructs when there is no single or simple way to describe reality from a limited perspective. Examples of this type of complementarity are particular-contingent, observed-observer, thought-action, mind-body, certainty-uncertainty, yin-yang, male-female, microcosm-macrocosm, and part-whole. Complementarity as a description for reality works because of underlying unity. In other words, complementarity validates the spiritual worldview of wholeness.

We can expand on a few of these complementary constructs from the standpoint of spiritual ideology.

- *Mind-body.* These are so intimately connected that nowadays psychologists consider them one. For a living organism, they must coexist.
- *Observer-observed.* Based on countless experiments with quantum particles, it is known that observation affects the observed. Observation (consciousness) influences matter precisely because matter is an epiphenomenon of consciousness. If consciousness were derived from matter, then logically it would be impossible for observation to affect physical reality.
- *Certainty-uncertainty.* Uncertainty exists in the physical and mental realms since everything in these realms has wave-like nature, and by definition a wave is nonlocalized with indefinite position. Additionally, sense organs and instruments have limited precision, and the very act of physically measuring or observing an object affects it. Certainty only exists in Cosmic Consciousness since it has no wave-like character and has no movement or vibration whatsoever in time or space.
- *Microcosm-macrocosm.* Complementarity of these constructs implies that whatever exists below also exists above and vice versa. For example, humans possess consciousness; therefore the universe is conscious. The unit mind reflects Cosmic Mind, etc.
- *Part-whole.* This describes the essence of the spiritual worldview. Ultimately, the creation is one, but it is perceived as consisting of parts because of a lack of cosmic perspective.

The nonlocality of mind

The bottom-up ontology of materialism that purports the equivalence of mind and brain is demonstrably false. Any attempt to explain numerous mental phenomena using materialistic models leads to impossible contradictions. On the other hand, downward causation, which postulates that mind is nonlocal and nonmaterial and thus subtler than brain, is consistent with myriad observations and a massive amount of experimental evidence. Below is a brief review of the evidence.

- *The unity of sensory experience.* There is no identifiable anatomical or brain basis that explains how sensory inputs are unified into a coherent experience. It can be concluded that mind, not brain physiology, is responsible for this.
- *Psychosomatic illness.* It is well known that mental states affect the body. Psychological feelings such as hopelessness and depression bring about increased risk of chronic disease, while feelings of joy and laughter improve health. Another example of mind affecting body is a placebo that produces a physiologic effect. The connection between mind and body is so strong that doctors sometimes call this "mind-body unity."
- *Other mind-body effects.* Examples include stigmata, localized skin responses such as blisters and skin writing, false pregnancy, whitening of hair or skin in response to severe fright or emotional stress, hypnotic effects on autonomic functions, allergies, and skin changes.
- *Memory and dreams.* Both are witnessed by mind. If memories were simply stored physiologically in the brain, then they could only be replayed and not witnessed from a third-person point of view. Mind, which is unitary and nonphysical, is able to supply this point of view.
- *Mystical experience.* The experience is universally described as entering into a clear, exalted state of ecstasy and unitary, limitless Consciousness. The similarity across cultural, religious, and national differences is indicative of a life-changing transcendental experience.
- *Reincarnation.* There is a preponderance of evidence that some people, especially children, have accurate memories of previous lives. Mind, which is nonphysical, survives death but carries memories and reactive momenta from one lifetime to the next. This also explains genius and other cases where a child spontaneously develops extraordinary abilities without any formal training.
- *Out-of-body experiences.* There have been thousands of cases reported of people floating above their body and witnessing events that took place from this unique perspective. Of particular interest are blind

persons having visual experiences and who are able to accurately describe events that took place while they were unconscious.
- *Near-death experiences.* There is no medical explanation of how people can experience vivid consciousness outside their body when they are clinically dead. Secondly, there is remarkable similarity between accounts, regardless of age, nationality, religion, race, culture, and other demographics. The fact that mind can function when the body and brain are "turned off" or considered clinically dead means that mind is nonlocal and separate from brain.
- *Extrasensory perception.* There is overwhelming scientific evidence that ESP is real. Positive results have been accumulating for over 125 years, indicating humans are capable of telepathy, clairvoyance, precognition, and psychokinesis. In most people, this capability is weak and the effect is shown by studies of many volunteers using statistics. However, for a few exceptional individuals a robust effect can be observed.
- *Mind affecting machines.* Scientists at the Princeton Engineering Anomalies Research Laboratory concluded after thirty years of study that intentions, emotions, and attitudes of human operators affect sophisticated equipment. Major world events have been shown to alter the output of random number generators. Studies of Intention Imprinted Electrical Devices indicate that a reproducible and robust affect is produced on various experimental conditions when a device is placed nearby with a specific intention imprinted upon it by an experienced meditator.

Can science prove the existence of God?

Science is primarily involved with proposing theories and testing them experimentally. Experimental evidence can validate a theory but proof in the sense of that obtained in mathematics can never be obtained. This is why such discoveries as evolution, relativity, and quantum mechanics are still labeled theories. This is despite the fact that countless experiments and observations have confirmed these theories. In addition, numerous predictions of these theories were subsequently shown to be correct. Therefore, in a sense science never offers proof of any hypothesis other than simple tautologies, such as electricity equals the movement of electrons or action begets reaction. There is always a possibility that some future experiment will contradict the current theory and require that it be amended. Therefore science is like quantum physics; it deals not in certainties but in likelihoods or probabilities. Hence, to say that science can prove the existence of God is illogical, since it cannot even prove the existence of the atom. However, this should not stop one from studying whether the existing data verifies the existence of Consciousness as the emergent property of creation.

We expect that any attempt to introduce the concept of God or Consciousness into science will be met with strong criticism by skeptics. There is good reason for this attitude. In the past when there were unexplained phenomena, they were often attributed to a higher being or god, and subsequently most of these unexplained phenomena were found to have natural explanations. Moreover, science has a long history of banging heads with religion; most scientists today are quite wary of studying anything that could be considered religious. Skeptics object to a paranormal explanation of unexplained phenomena by insisting that the experiments are flawed or that science will discover an explanation in the future. Furthermore, it can be argued that you cannot use spatial and temporal phenomena to prove the existence of something that transcends them. Other scientists may point out that science attempts to discover natural explanations for phenomena, but inviting God into the picture introduces the supernatural, which is not scientific. The problem with this argument is that science is unable to explain the space-time matter worlds of quantum physics using natural descriptions.

These arguments against theorizing and studying Consciousness as the foundation for reality are invalid. If science is in the business of studying and explaining phenomena in the natural world, and if Consciousness is the ground substance of the natural world, then science has to address this issue. Otherwise science could be regarded as incomplete—or worse, obsolete. We believe that there are sufficient verifiable indications for the existence of God—not in the Judeo-Christian sense, but as Consciousness—that for science to ignore the evidence and avoid studying the role that Consciousness plays in manifesting reality is to do a great disservice to humankind.

Although science cannot prove that God exists, there can be no doubt that consciousness exists. Rene Descartes is famously known for confirming this fact in his philosophical statement "I think, therefore I am." Although he had it backward, nonetheless conscious awareness is the experience of every human being. Furthermore, the factual existence of nonlocality in both the physical and mental spheres is proof positive that the bottom-up or materialistic worldview is false since it is founded on the premise that only local interactions can occur. The logical conclusion is that the top-down ontology of the spiritual worldview is correct. Does this constitute proof? The answer is no. But does the preponderance of evidence point to the existence of electrons, atoms, or water molecules? Yes, of course. Similarly, the preponderance of evidence points to Consciousness as the basic stuff of creation. Finally, if Consciousness is equated with God, then science and the scientific method could potentially prove that God exists through its study of Consciousness.

The one discovery that has put a dagger through the heart of materialism is entanglement. It says that particles can be on opposite sides of the universe, yet remain in intimate contact with one another, and this contact is not mediated by force fields or limited by the speed of light (local signals). We might wonder why this single most important discovery of science in the last one thousand years about the nature of reality has not shaken the apparent romance that scientists have today with materialism. The answer seems to be the same as why entanglement has not gained the imagination of the public. Even physicists have labeled the phenomena of entanglement as "weird," "crazy," "bizarre," and "inexplicable." Most scientists, to say nothing of the public, have yet to come to an understanding of what nonlocality implies about the nature of reality. Nonlocality implies wholeness, and such a concept is foreign to Western thinking. Neither science nor Western religions have introduced the concept of wholeness as a feature of reality. The common-sense experience of separateness is accepted as a fact of nature, and people are not encouraged by intellectuals, scientists, or theologians to consider the possibility that a deeper level of reality exists that connects everything and everybody. Hence it is no surprise that the ramifications of nonlocality are lost on the public and on most scientists.

Summary

Nonlocality is an experimentally verified quality of reality for both the physical and mental realms. Nonlocality negates the materialistic worldview but is entirely consistent with the spiritual worldview. Such a worldview explains the many paradoxes that have arisen in science today and provides a new model for how human society can progress in the future.

18
The Problem of the Day

> The entire humankind of the universe constitutes one singular people. All humanity is bound together; those who are apt to remain oblivious of this very simple truth, those who are prone to distort it, are the deadliest enemies of humanity. Today people should identify these foes very well and build up a healthy human society, totally ignoring all obstacles and difficulties. It must be borne in mind that so long as a magnificent, healthy and universalistic human society is not well established, humanity's entire culture, and civilization, its sacrifice, service and spiritual endeavor, shall not carry any worth whatsoever.
>
> — Shrii Shrii Anandamurti
> Ananda Vanii, 1973

TODAY HUMAN SOCIETY IS characterized by isolation, disconnectedness, and powerlessness. People have begun to distrust their own institutions and believe they are designed to advance the interests of a privileged few. People feel unworthy and dehumanized and the size of obstacles have become exaggerated. In such circumstances, some people respond to the powerlessness and separation they feel with pointless acts of violence.

Globalization and advances in technology have displaced workers. Some communities have become undermined by unemployment and the rapid movement of people. Family structure has disintegrated around the world. Radical, rapidly growing, self-destructive ideologies in the Middle East have destabilized the region. The global order has started to come undone as income inequality has increased and political ideologies have become more polarized, entrenched, and less willing to compromise.

Climate change brought on by the indiscriminate burning of fossil fuels and reliance on animal agriculture as a source of food threatens to displace millions of people, cause mass extinction of species, and rapidly alter the lands and waters that humankind depends upon for survival. In addition, there continues to be devastating insults to the environment by other human activities.

Then there is the skyrocketing cost of healthcare and the fact that close to 30 percent of scarce healthcare dollars in the US go to "end of life" medical services, many of which are used to prolong the life of terminally ill patients that have negligible quality of life. In addition, the US spends three times the amount of money per person on healthcare than Great Britain, and Americans pay the highest cost for prescription drugs.

Communalism, provincialism, and nationalism are still rampant worldwide and are sources of conflicts, wars, and terrorism. Many people have a greater loyalty to an ethnic, religious group, or state than to society in general, and probably outnumber those that have a universal or spiritual outlook.

Religious fundamentalism continues to propagate an ideology of separateness in its efforts to reestablish archaic religious dogmas that deny the divinity of humankind. Even the mainstream religious institutions in the West affirm the "reality" of the material world and do little to criticize the accumulation of excessive resources by a few individuals while other members of society suffer from want.

The great social institutions of liberal education, democracy, and capitalism are beginning to crumble under the weight of the dehumanizing materialistic worldview. The goal of a liberal education is to prepare individuals to deal with complexity, diversity, and change, and provide them with a broad knowledge of the wider world (e.g. science, culture, and society). However, by emphasizing the material and ignoring the spiritual, it is failing to inspire a sense of universalism, social responsibility, and meaning in students.

Today democracy is directly threatened by increasing inequality, as billionaires manipulate the media and finance candidates that will further their selfish interests. In the US, a corrupt campaign finance system allows a few very wealthy individuals with almost unlimited funds to essentially buy elections and exert disproportionate influence on the government and judiciary to the detriment of the general population. Inequality is probably the greatest present threat to democracy. It is the cause of increasing social polarization, and it is destroying people's trust in democratic institutions.

Capitalism, the cherished economic system of the West, which is credited with bringing untold prosperity and social freedom, is beginning to crumble. Capitalism depends on a robust middle class to supply the capital and labor needed for economic growth, but growing income inequality is undermining this. Furthermore, an increasing world population and diminishing resources threaten to attenuate the rampant consumerism that is required to support this economic system. With a deep understanding and acceptance of spiritual ideology, a day will come when both capitalism and communism (state capitalism) disappear because they rely exclusively on selfish, profit-based, material incentives and lead to the accumulation of wealth for the benefit of a few rather than for the welfare of all. Being devoid of spiritual principles, capitalism is not congenial for the integrated growth of human society.

Finally, human society is plagued by terrorism, mass murder, drug addiction, suicide, and numerous other antisocial activities. People are either not concerned with or unaware that their actions might result in consequences in the future. A lack of knowledge and understanding of the Law of Karma contributes to this scourge. Furthermore, Western religions no longer strongly discourage such activities because they teach the implausible doctrine that committing sin will result in eternal separation from God. In the name of religion, extremist ideologies have taken root that preach that killing innocent people in the name of furthering their fundamentalist cause will send them to heaven.

Materialist doctrine is the root cause of society's problems

All the problems outlined above can be traced directly or indirectly to the general acceptance of the materialistic worldview by a large majority of scientists, intellectuals, politicians, economists, capitalists, and the public in general. This worldview degrades the human mind by assigning a firm reality to the physical world and promotes an ideology of separateness. It denies people's connection to the Divine and discourages them from diving deep within to uncover the source of their being. It is a false ideology, and it needs to be supplanted by a new philosophic model of reality based on wholeness, if human society is to progress.

Matter is the be-all and end-all of materialism. Materialism proclaims that mind has been created out of matter by a process of chemical transformation and thus does not have any independent or special significance beyond its materialistic value. This implies that the mind has no business making moral or other judgments since its very existence is denied. The search for meaning in life is a waste of time, and being awestruck by beautiful scenery or a colorful sunset is a physiological response to the environment. Love is merely a condition brought on by chemical processes in the body, an offshoot of the need for the human species to procreate.

Materialism proclaims that there is only one life to live and there is no reason to be concerned that consequences of actions performed in this lifetime could haunt one in a future life. Because there is a disconnect between personal actions and their consequences, people are unaware that antisocial behavior is not in their own self-interest. They do not feel that there is any debt to be paid if they act against the interest of others or society in general.

Materialism functions by imposing social pressure on people to enjoy material objects. It promotes consumerism, which has quietly become the principle way in which many people measure themselves and others. It is driven by the creation and encouragement of desire for material well-being as it idealizes an artificial lifestyle that is promoted as the principle means of achieving happiness and the good life. For those caught in its web, energies are directed toward the attainment of wealth, possessions, and social appearances, instead of the more gratifying pursuit of personal growth and being. The illusion of consumerism is created largely by commercial interests and the entertainment industry. The emphasis on having rather than being is one of the factors that is currently degrading human society since it leads inexorably to selfish material desires rather than a concern for the well-being of society. Unfortunately, Western religions do little to diminish the discontent among the have not's in their aspiration to emulate the haves.

Materialism fosters individualism at the expense of self-sacrifice and service to others. From a biological and evolutionary perspective, there is nothing to be gained personally by service to others outside one's immediate family. Individualism serves to isolate people and inhibits feelings of family and friendship for other members of society. As a result, people are unable to harmonize their diverse ideas and ideologies, and progress together, gradually transforming their self-interests into a unified rhythm, which is characteristic of a healthy human society.

Materialist doctrine considers the body to be a material object, and thus great emphasis is placed on allopathic medicine. The cost savings of alternative medical approaches such as homeopathy, acupuncture, biofeedback, meditation, prayer, chiropractic, naturopathy, Christian Science, and Ayurveda are not realized. The medical establishment has been schooled in reductionist ideology and for the most part still believes that alternate approaches to healthcare have little, if any value. As a result, integrated medical treatments are ignored and a greater emphasis is placed on treating disease as opposed to preventing it.

Scientific materialism has been at war with religion for over five hundred years. The founding fathers of the US were keenly aware of the threat that religious zealots posed for the young republic and were careful to allow people to practice their religion freely while insuring separation of church and state. As a result, secular and materialistic ideologies are freely taught in the public schools while any teaching of spirituality is taboo since it could be construed as religious doctrine. Materialism cannot be blamed for this situation, but there is little doubt that spiritual values and ideas will not be taught in schools until science adopts a correct theory of reality that incorporates Consciousness. At the same time, people need to be better educated about what spirituality is, making it clear that it is not religion.

Today scientific materialism has largely won the hearts and minds of most people living in the developed world. Many people who as children went to church regularly with their parents have since left the church and are now skeptical that God exists. As a result, they have gravitated toward the prevailing scientific ontology, in which everything is derived from matter and conscious existence ends when the body dies. In its fight with religion to win over the hearts and minds of people, science has emerged the clear victor. Most of the respected scientists, doctors, intellectuals, business, and political leaders in the West subscribe to the doctrine of materialism. However, materialism is no less dogmatic than religious fundamentalism. Both are faith based. For religion it is the belief that scripture is the authoritative truth, while for science it is the belief that the natural world is derived from matter and that the bottom-up theory of reality must be true—after all it has worked so well in the past and explains most everything. However, the problem with dogma is that it is a preconceived idea that forbids human beings to go beyond its limits. In this situation, the human intellect cannot freely function, and intellect is one of the greatest treasures of human beings. This is a tragic situation since dogma prevents the liberation of human intellect.

The Islamic jihadist movement that is destabilizing the Middle East is largely a response to the materialist ideology of the West. Those individuals that are attracted to such destructive and inhuman ideologies are fearful of the growing secularism that threatens to sweep away the religious fundamentalism that they believe in. Change in this difficult situation will not come by applying military pressure from without but from within through better education and pressure from more moderate Muslims.

19
The Spiritual Worldview to the Rescue

The movement and the path, the means and the chariot are all inseparably linked. The path is not always easily accessible, smooth and littered with flower petals; nor is it always inaccessible, thorny and covered with stones. One must keep one's eye fixed on the Goal. This Goal provides inspiration, supplies the means for forward movement, and makes the little lamps of life infinitely effulgent. Since eternity this very Goal has provided and is providing inspiration to all and will continue to do so in future; and by revitalizing the life-force as if with a flow of water, it will make the earth ever full of sweetness, and at the same time it will keep the triumphant flag of humanity flying on top of the golden mountain peak. So let one's vision be fixed on the Goal. There is no necessity to think of anything else.

— Shrii Shrii Anandamurti
Ananda Vanii, 1987

Social changes

SPIRITUAL IDEOLOGY PROCLAIMS THAT every person, as well as every object in this universe, is a manifestation of the Cosmic Entity. Human society is one, and each individual should be considered a member of a large family. In a joint family, every member is provided with the necessary food, shelter, clothing, education, and medical care needed for their survival, according to the economic resources of the family. If any member accumulates resources in such a way that it harms the other family members then they are chastised. On the other hand, if a family member is unable to contribute their fair share to the family because of mental or physical difficulties then the family will still accept that person with open arms and attempt to help them in any way they can. Human society must strive to do the same.

Many have tried to jeopardize the unity of the human race by creating factions. Such persons have a stake in creating divisions; they survive on the mental weaknesses of people and on their dissensions. Such people are afraid of the spread of an ideology of wholeness and exhibit their intolerance toward it in numerous ways, such as creating divisiveness, false propaganda, and lies. Knowledgeable people are not influenced by the dogma of separateness; they continually strive to perceive the unity of everything and everybody and think and act according to universalism and a new version of humanism, which are central features of a spiritual worldview.

The propagation of this new humanism or neo-humanistic ideology will help solve one of the greatest threats to humankind today—anthropogenic climate change. Neo-humanism purports that all life is sacred, not just human life. Human life should be considered more valuable than animal life, but if there is no need to take animal life in order to survive and be healthy then it is morally indefensible to kill animals for food.

According to a 2006 report published by the United Nations Food and Agriculture Organization, and a 2009 report by the Livestock and Climate Change (LCC) environmental assessment experts at the World Bank, animal agriculture is responsible for over half of the total global greenhouse gas emissions.[1,2] This means that the unnecessary and unsustainable consumption of meat in developed countries accounts for more greenhouse gas emissions than the transportation, electrical generation, and other industrial uses of fossil fuels combined. It is estimated that the feeding of livestock now uses over 30 per cent of the earth's entire arable land surface, mostly in producing feed for the animals. Animal agriculture also drives deforestation, especially in South America where approximately 70 percent of former forests in the Amazon have been turned over to grazing. In the US, approximately 90 percent of the food grains that are grown are fed to livestock. Cattle also cause widespread land degradation through overgrazing, compaction, and erosion. Animal agriculture is also damaging to the environment because of its use of enormous amounts of scarce water resources and the pollution caused by animal waste, fertilizers, and pesticides. The adoption of spiritual ideology by a majority of the world's population would greatly reduce the misallocation of scarce resources for the feeding of animals, lower greenhouse gas emissions, and improve the health of individuals. Additionally, as spiritual values take hold, the powerful antienvironmental interest groups will lose their political grip, allowing renewable energy use to grow and replace more fossil fuels.

The spiritual worldview teaches that Consciousness has no beginning or end. Human beings possess consciousness and this aspect of their being is eternal, unless or until it is merged with Cosmic Consciousness. In order to achieve such perfection many lifetimes may be necessary; thus there is life after death and one must suffer the consequences of one's actions—both good and bad. A simple understanding of reincarnation and the Law of Karma by most of the world's population would go a long way toward establishing a new understanding of the human condition. People would be more motivated to act in their true self-interest and discover the benefits of altruism, selfless service, and the sharing of scarce resources for the benefit of all.

Additionally, a new model for end-of-life care would emerge if people realized that death is a transition and not final. Instead of spending a third of limited healthcare resources on prolonging the life of terminally ill patients with questionable quality of life, loved ones and the medical establishment would be more open to letting people leave their body when they are ready.

Other aspects of the industry would also change with the realization that the body has both physical and psychic strata, and that some illnesses are better treated using alternative and integrated approaches instead of the current body-is-machine mentality of allopathic medicine. The alternative approach of mind-body medical treatments is both cost effective and often can affect a real cure instead of just treating the symptoms. Moreover, as more people gravitate toward the path of the seeker they naturally adopt a more healthy lifestyle. They suffer fewer chronic illnesses. The scourge of obesity will be reduced. Universal healthcare will become more affordable and more emphasis will be put on disease prevention as opposed to treatment.

When people begin to realize that happiness comes from within rather than from without, they naturally feel less driven to acquire material wealth and possessions. More emphasis will be put on being as opposed to having. The sickness of accumulation of wealth, power, and fame will gradually be replaced by a healthy appreciation for the inner treasures obtained through spiritual practices such as meditation. Music that lifts the soul, such as kirtans and bhajans, will increase in popularity and spiritual communities that function as cooperatives will flourish. The entertainment industry will also change as people gravitate toward spiritually uplifting art and other media as opposed to tales of violence and sex. As people develop awareness of the meaning and purpose of life, they will start to look up to spiritually elevated individuals. Individuals that acquired vast wealth or power by taking advantage of a flawed system or by exploiting people will no longer be idolized, and those with great physical prowess will no longer be held in higher esteem or be paid more than those persons who display great intellectual ability.

Economic changes

Spiritual ideology is inconsistent with the matter-based socioeconomic systems of capitalism and communism. A new socioeconomic system based on the welfare of the many as opposed to the few will emerge. One such system is that of P. R. Sarkar, called Progressive Utilization Theory or PROUT. PROUT offers a panacea for the integrated progress of human society. It aims to bring about equilibrium and equipoise in all aspects of socioeconomic life by totally restructuring economics. Sarkar positioned it as an alternative to communism and capitalism. It recognizes all material goods as common to all people and seeks the rational and equitable distribution of physical resources to maximize the physical, mental, and spiritual development of humankind. The PROUT system of economic development seeks to guarantee everyone the five minimum requirements of life—food, clothing, shelter, education, and medical care. As an incentive, surplus physical resources are distributed to people who best serve and make the greatest contributions to society. PROUT is a type of progressive socialism that would restrict the accumulation of excessive wealth by a few individuals to the detriment of the many, advance cooperatives as the model for most nonessential manufacturing, and still incentivize people to be creative and productive. A new socioeconomic system based on spiritual principles will go a long way toward integrated progress in the economic sphere and the creation of a healthy human society.

Spiritual ideology encourages cooperation and service in place of the selfish individualism that is so destructive to civilization. It teaches that service is good for the soul as well as for the person served. The service that most effectively dissolves the ego is that which is performed out of the goodness of one's heart, requiring nothing in return, not even recognition or thanks. Such service can be physical service, such as helping to build a shelter, or attending the sick; or economic service, such as relief work, feeding the poor, or helping the needy find a job. A higher form of service is intellectual service, such as teaching skills, general knowledge, morality, and spiritual philosophy. The highest form of service is spiritual service—performing sincere spiritual practices. Such practice helps the individual attain happiness, wisdom, empathy, and eventual unity. Others will want to emulate them due to their influence and example, and will thus be drawn toward the spiritual path. Unlike physical forms of service, the effects of intellectual and spiritual service can be permanent in nature.

The material needs of people and other living organisms must be satisfied to maintain life, and life must be maintained to reach God. All forms of true service weaken the grip of ego and bring one closer to the goal of unity. Hence service advances one's personal growth and produces a more harmonious human society.

Knowledge of spiritual ideology will inspire more people to perform spiritual practices. Such practices produce a love for God. All great saints and prophets had an overwhelming love for God, and they radiated this love. Naturally their charisma attracted many followers. Where there is love for the Cosmic Entity there is no personal ego, since ego is only involved with the attachments for finite things. Devotional love is directed inward, as opposed to love for a spouse or child, which is directed outward. Arousing and attaining love for God is not easy. One can spend their life tending to the poor, attending daily church services, reciting prayers or mantras, undergoing long fasts, or performing extreme acts of self-denial, and never obtain more than momentary feelings of devotion for God. It helps to live by the Golden Rule and to try to love and respect all manifestations of God, but devotion is a state of mind, not a practice. This blessed state is normally obtained by practicing meditation, kirtan, surrender, and selfless service. Devotional love for God creates complete trust in him and the assurance that all actions one performs are performed by him. Those persons who can develop strong love of God are well on their way to becoming a seer and are invaluable for inspiring others to enter the path of knowledge. The contributions that such people make to society are priceless.

Strong scientific evidence for the top-down ontology of spiritual ideology already exists. However, if even a fraction of the resources put into discovering new subatomic particles, cancer research, or space exploration were put into the scientific research of Consciousness, then rather quickly the evidence would become overwhelming in favor of this worldview. A new paradigm of science would emerge that puts Consciousness in its rightful place as the ground substance of creation. The philosophical concept of wholeness would be incorporated into science in this new paradigm. The dogma of scientific materialism would finally be demolished. Except for a few remaining religious fundamentalists, the vast majority of people on earth would have the intellectual freedom to explore spiritual ideology and learn of the great promise that it offers the human race. This change will eventually take place and will not only revolutionize science but also education, philosophy, ethics, and theology. Human society will enter a glorious new phase of development distinguished by harmony, peace, economic prosperity, political unity, and individual spiritual growth.

The final journey

Spiritual ideology implies that what is true above is also true below. In other words, the microcosm reflects the macrocosm. Ultimately they are the same; however, we live in a relative reality that demands our attention. There is a Sanskrit aphorism that describes this: *Brahma satyam jagadapi satyamapeksikam.* "Brahma is absolute truth and the universe is relative truth." Although the Cosmic Entity is unchanging and infinite, when it manifests as the creation it comes into the realm of relativity under the bondages of time, place, and person. The microcosm is under the same bondages and is a relative entity that undergoes constant change.

Since the cosmos is perceived as relative reality, it appears to be a relative truth to the changeable individual. Hence one cannot completely escape this relative reality. It is a mistake to try to deny this world by calling it an illusion and saying that only God is real. One has to deal with this world in order to transcend it.

Everything in the material world is formed from Consciousness and human beings possess a developed mind and self-awareness; they have an underlying thirst for limitlessness. The attraction for the Great steers them toward the Cosmic Entity. This Entity is the source of all being and one must ultimately return to it in order to fulfill one's ultimate destiny. Every human being is on an endless quest for the Great.

Ultimately, the unity that one seeks is attained by surrender. First the attachment to crude material objects is lost. Next one's own individuality is surrendered. Oneness is experienced when the little "I" is surrendered completely to the "Great I" and nothing is held back. Such self-surrender is not akin to suicide. On the contrary, the individual soul will have its full expression. Its existence does not become contracted but enlarged as it realizes its true being as Cosmic Consciousness.

The process is called enlightenment because it is awakening to a clear understanding of the wholeness that is the Ultimate Reality. The word "awakening" is useful because it is analogous to awakening from a dream. While dreaming, one does not question the reality of the images that are witnessed, even though they may be quite bizarre. Upon waking up, one immediately dismisses the dream as unreal. Similarly, the illusion of separateness or nonunity appears real, but upon awakening to an enlightened state, one knows that the waking state of consciousness was illusory.

In this awakened state, one loses the sense of a separate selfhood. It is the state of consciousness experienced by the seer. Everything is seen and felt as God and there is an end to suffering. A natural state of bliss arises from the nonseparateness that occurs when one lives totally in the here and now of pure being. One continues to experience life with its joys, pleasures, pain, and love, but these experiences are not resisted by the illusory "me." In other words, one's experiences are witnessed by the mind but have no effect on the "I that is."

Readers may be convinced that this ideology of the future is logical and correct in its description of reality but should realize that an intellectual understanding of reality does not bring more than intellectual satisfaction. In order to experience the ecstasy of oneness and unlimited happiness that exists in the infinite, unqualified Cosmic Consciousness, one must begin the difficult journey of self-discovery. Real change in human society can only take place in the individuals that make up that society.

Notes

Introduction

1. Bernard d'Espagnat, *In Search of Reality* (New York: Springer-Verlag, 1983).
2. See for example: Richard Dawkins, *The God Delusion* (New York: Mariner, 2008); Sam Harris, *Waking Up: A Guide to Spirituality without Religion* (New York: Simon & Shuster, 2014); Christopher Hitchens, *God is Not Great: How Religion Poisons Everything* (New York: Hachette, 2007).

Chapter 1

1. Stephen Hawking and Roger Penrose, *The Nature of Space and Time* (Princeton: Princeton University Press, 1996) 89-90.

Chapter 3

1. Werner Heisenberg, *Physics and Philosophy: The Revolution in Modern Science* (New York: Harper & Row, 1962) 71 and 191.

Part II

1. Edward F. Kelly and Emily W. Kelly, *Irreducible Mind: Toward a Psychology for the 21st Century* (Lanham, MD: Rowman & Littlefield, 2007) 117.

Chapter 5

1. Von der Malsburg, "Binding in Models of Perception and Brain Function," *Current Opinion in Neurobiology*, **5**, (1995): 520-526.
2. J. R. Searle, "Minds, Brains, and Programs," *Behavioral and Brain Sciences*, **3**, (1980): 417-24.
3. Edward F. Kelly and Emily W. Kelly, *Irreducible Mind: Toward a Psychology for the 21st Century*, 129
4. R.L. Moody, "Bodily Changes during Abreaction," Lancet, 1, (1948): 964.

5. Edward F. Kelly and Emily W. Kelly, *Irreducible Mind: Toward a Psychology for the 21ˢᵗ Century*, 196-8.
6. Ibid p. 216.
7. L.K. Kothari, Arun Bordia, and O. P. Gupta, "The Yogic Claim of Voluntary Control over the Heart Beat: an Unusual Demonstration." *J. American Heart Assoc.*, **86**, no. 2, (1973): 284.
8. Ian Stevenson, *Telepathic Impressions: A Review and Report of 35 New Cases* (Charlottesville, VA: University Press, 1970).
9. Edward F. Kelly and Emily W. Kelly, *Irreducible Mind: Toward a Psychology for the 21ˢᵗ Century*, 172-3.
10. Larry Dossey, M.D. Healing Words: The Power of Prayer and the Practice of Medicine (New York: HarperCollins, 1993).
11. Richard Gerrig and Phillip Zimbardo *Psychology and Life*, 20th edition (Essex England: Pearson Education, 2014).
12. Rick Hanson and Richard Mendius, *Buddha's Brain: The Practical Neuroscience of Happiness, Love and Wisdom* (Oakland, CA, New Harbinger, 2009) 7.
13. Edward F. Kelly and Emily W. Kelly, *Irreducible Mind: Toward a Psychology for the 21ˢᵗ Century*, 251.

Chapter 6

1. Robert Ullman and Judyth Reichenberg-Ullman, *Mystics, Masters, Saints, and Sages, Stories of Enlightenment* (Berkeley, CA: Conari Press, 2001) 37.
2. Gopi Krishna, *Living with Kundalini* (Boston: Shambhala, 1993) 2.

Chapter 7

1. Bruce Leininger and Andrea Leininger, *Soul Survivor: The Reincarnation of a World War II Fighter Pilot* (New York: Grand Central, 2009).
2. NBC Nightly News on March 20, 2015.
3. Jim B. Tucker, *Return to Life: Extraordinary Cases of Children who Remember Past Lives* (New York: St. Martin's, 2013) 88-119.
4. Ian Stevenson, *Children Who Remember Previous Lives: A Question of Reincarnation* (Jefferson, NC: McFarland, 2001).
5. Jim B. Tucker, *Life before Life: Children's Memories of Previous Lives* (New York: St. Martin's, 2005).
6. Ian Stevenson, *Reincarnation and Biology: A Contribution to the Etiology of Birthmarks and Birth Defects* (New York: Praeger, 1997).
7. Edward F. Kelly and Emily W. Kelly, *Irreducible Mind: Toward a Psychology for the 21ˢᵗ Century*, 233.
8. Shirley MacLaine, *Out on a Limb* (New York: Bantam, 1983).
9. Raymond A. Moody, Jr., *Coming Back: a Psychiatrist Explores Past-Life Journeys*

(New York: Bantam, 1995).
10. Ibid. 34-45.
11. Ibid. 51-79.
12. Brian L. Weiss, *Many Lives, Many Masters* (New York: Simon & Schuster, 1988).
13. A. T. Mann, *The Elements of Reincarnation*, (Rockport, MA: Element, 1995).
14. Ian Stevenson, *Children Who Remember Previous Lives: A Question of Reincarnation* (Jefferson, NC: McFarland, 2001) 36.
15. Herbert B. Puryear, *Why Jesus Taught Reincarnation* (Scottsdale, AZ: New Paradigm Press, 1993).
16. Quincy Howe, Jr., *Reincarnation for the Christian* (Philadelphia: Westminster Press, 1974).
17. Elaine Pagels, *The Gnostic Gospels* (New York: Random House, 1979).
18. Matthew 11: 13-14; Matthew 17: 10-13.
19. Matthew 16:3; Luke 9:18; Mark 8:26.
20. Ian Stevenson, *Children Who Remember Previous Lives: A Question of Reincarnation*, 29.
21. Ibid. 30.

Chapter 8

1. Larry Dossey, *Recovering the Soul: a Scientific and Spiritual Search* (New York: Bantam, 1989) 17-19.
2. H. Hart, "ESP Projection: Spontaneous Cases and the Experimental Method," *Journal of the American Society for Psychical Research*, 48, (1954) 121-46.
3. Sam Parnia, *Erasing Death: The Science that is Rewriting the Boundaries between Life and Death* (New York: HarperCollins, 2014).
4. Robert Bruce and Brian Mercer, *Mastering Astral Projection: 90-day Guide to Mastering Out-of-body Experience* (St. Paul, MN: Llewellyn, 2004).
5. These include the Monroe Institute's Nancy Penn Center in Virginia, the Center for Higher Studies of Consciousness in Brazil, the International Academy of Consciousness in southern Portugal, which features the Projectarium, a spherical structure used exclusively for practice and research on out-of-body experience, and Olaf Blanke's Laboratory of Cognitive Neuroscience in Switzerland.
6. Edward F. Kelly and Emily W. Kelly, *Irreducible Mind: Toward a Psychology for the 21st Century*, 370-3.
7. Raymond A. Moody, Jr., *Life after Life: The Investigation of a Phenomenon—Survival of Bodily Death* (New York: HarperCollins, 1975).
8. Jeffrey Long and Paul Perry, *Evidence of the Afterlife: The Science of Near-Death Experiences* (New York: HarperCollins, 2010).
9. Edward F. Kelly and Emily W. Kelly, *Irreducible Mind: Toward a Psychology for the 21st Century*, 418.

10. Ibid. 419.
11. Raymond Moody, Intelligence Squared debate entitled "Death is not Final." See: http://intelligencesquaredus.org/images/debates/past/transcripts/050714%20Death%20Not%20Final.pdf.

Chapter 9

1. Sean Carrol, Discover Magazine Blogs: http://blogs.discovermagazine.com/cosmicvariance/2008/02/14/american-association-for-the-advancement-of-pseudoscience.
2. Sean Carrol, Intelligence Squared debate entitled "Death is not Final." http://intelligencesquaredus.org/images/debates/past/transcripts/050714%20Death%20Not%20Final.pdf.
3. Literally an analysis of analyses. The aim is to derive a statistical analysis from a pool of similar studies. This allows for combination of information leading to higher statistical certainty than is possible from any individual study.
4. Dean Radin, *The Conscious Universe: The Scientific Truth of Psychic Phenomena* (New York: HarperCollins, 1997) 79-80.
5. Ibid. 87-89.
6. Guy L. Playfair, *Twin Telepathy* (Guildford, UK: White Crow, 2011).
7. Dean Radin, *The Conscious Universe: The Scientific Truth of Psychic Phenomena*, 99-100.
8. J. G. Pratt, et al, *Extrasensory Perception after Sixty Years* (Boston: Bruce Humphries, 1966) 42.
9. Dean Radin, *The Conscious Universe: The Scientific Truth of Psychic Phenomena*, 120.
10. Ibid. 126-130.
11. Ibid. 143.
12. Hardware RNGs are preferred and may employ radioactive decay or electronic noise. Noise spikes or decay signals are measured against an electronic oscillator such as a quartz crystal to produce thousands of spikes per second, which can be translated into a truly random sequence of zeros and ones that are accurately recorded by computer.
13. Dean Radin, *The Conscious Universe: The Scientific Truth of Psychic Phenomena*, 151.
14. B. Dunne and R. Jahn, "Experiments in Remote Human/Machine Interaction, *Journal of Statistics Edu.*, **6** (1992): 311-32.
15. Dean Radin, *Entangled Minds: Extrasensory Experiences in a Quantum Reality* (New York: Paraview, 2006) 195-202.
16. Dean Radin, *The Conscious Universe: The Scientific Truth of Psychic Phenomena*, 81-2.
17. Ibid. 80-1.
18. Ibid. 100.

19. Ibid. 139.
20. See for example: http://www.princeton.edu/~pear/
21. William A. Tiller., Walter Dibble, and Michael Kohane, *Conscious Acts of Creation: The Emergence of a New Physics* (Walnut Creek, CA: Pavior Publishing, 2001) 1-13.
22. A gas discharge tube is a bulb or tube (usually glass) with two or more electrodes that has been evacuated and filled with a gas or gas mixture, usually at somewhat less than atmospheric pressure. As the voltage applied across the electrodes is increased, there comes a point called the breakdown voltage at which ionization of the gas will initiate an avalanche process that spreads through the tube. The voltage at which breakdown occurs depends on various factors, such as gas composition, pressure, spacing of the electrodes, etc.
23. William A. Tiller., Walter Dibble, and Michael Kohane, *Conscious Acts of Creation: The Emergence of a New Physics*, 2-5.
24. Dean Radin, *Entangled Minds: Extrasensory Experiences in a Quantum Reality*, 136-141.
25. Gary E. Schwartz, *The Sacred Promise: How Science is Discovering Spirit's Collaboration with us in our Daily Lives* (New York: Atria Books, 2011).
26. H. E. Puthoff and R. Targ, "A Perceptual Channel for information transfer over Kilometer Distances: Historical Perspective and Recent Research," *Proceeding of the Institute of Electrical and Electronic Engineers*, **64** (1976).
27. Dean Radin, *The Conscious Universe: The Scientific Truth of Psychic Phenomena*, 105.
28. Sheila Ostrander and Lynn Schroeder, *Psychic Discoveries behind the Iron Curtain* (New York: Prentice-Hall, 1970).
29. Jess Stern, *Edgar Cayce the Sleeping Prophet: The Life, Prophecies, and Readings of America's Most Famous Mystic* (New York: Bantam, 1967).
30. Dean Radin, *Supernormal: Science, Yoga, and the Evidence for Extraordinary Psychic Abilities* (New York: Chopra Books, 2013).
31. Larry Dossey, *Recovering the Soul: a Scientific and Spiritual Search* (New York: Bantam, 1989), 113.
32. Maryann Mott, *National Geographic News*. January 4, 2005.
33. Don Oldenburg, *Washington Post*. January 8, 2005.
34. Maryann Mott, *National Geographic News*. November 11, 2003.
35. Dean Radin, *The Conscious Universe: The Scientific Truth of Psychic Phenomena*, 275.

Chapter 10

1. For example, see the books of Fritjof Capra (*The Tao of Physics* and *The Web of Life*); Gary Zukav (*The Dancing Wu Li Masters*); Amit Goswami (*The Self-Aware Universe, Quantum Creativity, Physics of the Soul, God is Not Dead, How Quantum*

Activism Can Save Civilization, The Visionary Window); Robert Nadeau and Menas Kafatos (*The Conscious Universe* and *The Non-local Universe*).

2. For example, many new stars can be observed forming in the Crab Nebula of the constellation Orion from the remnants of a supernova that was observed there by Arab astronomers in 1054 AD.

Chapter 11

1. Carl Jung, *Man and His Symbols* (New York: Bantam, 1964).
2. Joseph Campbell, *The Hero with a Thousand Faces*, 3rd Ed. (Novato, CA: New World Library, 2008).
3. Rupert Sheldrake, *A New Science of Life: The Hypothesis of Morphic Resonance* (Los Angeles: J. P. Tarcher, 1981).
4. Bernard d'Espagnat, "The Quantum Theory of Relativity," *Scientific American*, Nov. 1, 1979.

Chapter 12

1. Richard Dawkins, *The Blind Watchmaker: Why the Evidence of Evolution Reveals a Universe without Design* (New York: Norton, 1986).
2. Michael J. Behe, *Darwin's Black Box: The Biochemical Challenge to Evolution* (New York: Free Press, 1996).
3. For a comprehensible discussion of this problem with material Darwinism see: Thomas Nagel, *Mind and Cosmos: Why the Materialist neo-Darwinian Conception of nature is Almost Certainly False* (Oxford: Oxford University Press, 2012).
4. Amit Goswami, *Creative Evolution* (Wheaton, IL: Theosophical Publishing House, 2008).
5. Stephen Hawking and Roger Penrose, *The Nature of Space and Time*
6. (Princeton: Princeton University Press, 1996) 89-90.
7. Roger Penrose, *The Road to Reality: A Complete Guide to the Laws of the Universe* (New York: Knopf, 2005) 726-756.
8. Martin J. Rees, *Just Six Numbers: The Deep Forces That Shape the Universe* (New York: Basic Books, 2000).
9. Conscious intent in this case does not mean that individuals must be aware of the intent. For example, most of the manifestations of reality, such as the formation of a rain cloud do not depend on anyone's conscious desire or awareness but that of Cosmic Mind. Similarly, autonomic activities in the body do not depend on the conscious mind being involved but on subtler layers of mind/body, that is, the psychic body that reflects Cosmic Mind

Chapter 13

1. Max Planck quoted in *Accent Magazine*, "The Hidden World of Mind," Oct. 1972, New Delhi, India.
2. Wolfgang Pauli, *Writing on Physics and Philosophy* (Berlin: Springer- Verlag, 1994), 259
3. Ken Wilber, *Quantum Questions, Mystical Writings of the World's Greatest Physicists* (Boston: Shambhala, 1984) 173
4. Ibid., 175.
5. Ibid., 144.
6. Ibid., 150.
7. Erwin Schrödinger, *What is Life* (Cambridge: Cambridge University Press, 1947) 137.
8. Erwin Schrödinger, *My View of the World* (Woodbridge, CT: Ox Bow Press, 1983) 31-4.
9. Ken Wilber, *Quantum Questions, Mystical Writings of the World's Greatest Physicists*, 87.
10. Werner Heisenberg, *Across Frontiers* (Woodbridge, CT: Ox Bow Press, 1990) 105-6.
11. Werner Heisenberg, *Physics and Philosophy: The Revolution in Modern Science* (New York: Harper & Row, 1962) 71.
12. David Bohm, *Wholeness and the Implicate Order* (New York: Routledge & Keagan Paul, 1983) 175.
13. Ibid. 210.
14. Larry Dossey, *Recovering the Soul: a Scientific and Spiritual Search*, 175.
15. Henry Margenau, *The Miracle of Existence* (Woodbridge, CT: Ox Bow, 1984) 96.
16. Albert Einstein quoted in the *New York Post*, November 28, 1972.
17. Albert. Einstein, *Ideas and Opinions* (New York: Three Rivers Press, 1954) 12.
18. A list of a few of these modern-day visionaries would include Joseph Campbell, Fritjof Capra, Gary Zukav, Richard Alpert, Thomas Moore, Deepak Chopra, Larry Dossey, Eckhart Tolle, Dean Radin, Amit Goswami, Menas Kafatos, Robert Nadeau, Norman Friedman, Ervin Laszlo, and Ken Wilber.

Chapter 15

1. Britta K. Hölzel, James Carmody, Mark Vangel, Christina Congleton, Sita M. Yerramsetti, Tim Gard, and Sara W. Lazara, "Mindfulness Practice Leads to Increases in Regional Brain Gray Matter Density," *Psychiatry Research: Neuroimaging*, **191**, no. 1 (2011): 36–43.

Chapter 16

1. Carl G. Jung, *Modern Man in Search of a Soul* (New York: Harcourt Brace Jovanovich, 1933).
2. Deepak Chopra, *The Way of the Wizard* (New York: Harmony Books, 1995) 161.
3. Luke 18:17.
4. Mark 10:24, Matthew 5:44.
5. Matthew 5:39.

Chapter 19

1. FAO, 2006. "Livestock's Long Shadow: Environmental Issues and Options," Food and Agriculture Organization of the United Nations.
2. Goodland, R. and Anhang, J., "Livestock and Climate Change: What if the Key Actors in Climate Change were Pigs, Chickens and Cows?" (2009). *Worldwatch*, November/December 2009, Worldwatch Institute, 10–19.

Index

A

acharya 172
acupuncture 193
aerial factor 100, 103, 107
ajina chakra 118
akasha 99, 100
amygdala 164
anahata chakra 118
ananda 62, 108
Ananda Marga 97
Anandamurti, Shrii Shrii, 97, 189, 195
animal agriculture 190, 196
animal instincts 115, 133
annamaya kosha 113
anomalies 6, 14, 17, 47, 49, 87, 90, 95
anthropic principle 15, 134, 136
archetypes 115, 143
arrow of time 39, 43
asanas 60, 109, 119, 152, 162
Ashtanga Yoga 60
Aspect, Alain, 19, 20
Aspect experiment 20
astral projection 71
atman 108, 116
avidya 108, 109, 126, 157, 159
avidya Tantric 126
AWARE studies 71

B

Baha'u'llah 60
Behe, Michael, 207
Bell's theorem 19, 20, 179
Berkeley, George, 146
Bhagavad Gita 59
bhajans 198
Big Bang 11-14, 20, 37, 40, 97, 100, 134-136
Big Crunch 13, 15
biofeedback 52, 193
black-body radiation 17, 143
black hole 34, 41, 182
block time 33, 39
bodiless mind 88, 121-123, 126, 127, 156
Bohm, David, 147, 148, 208
Bohr, Neils, 17-19, 149, 150
bowerbird 133
Brahma 7, 98, 111, 130, 150, 152, 200
Brahma Chakra 98, 111, 130
Buddha 60, 89, 105, 126, 156, 203
buddhi 105
Buddhism 7, 67, 115, 162

C

Cambrian Period 132
Campbell, Joseph, 116, 207, 208
capitalism 190, 191, 198
Capra, Fritjof, 207, 208
Cayce, Edgar, 89, 206
chakra 118, 119, 152, 165
chanting 162
chi 119
chiropractic 193
Chopra, Deepak, 172, 206, 208, 209
Christ 52
Christianity 67, 151
Christian Science 193

clairvoyance 71, 78, 80, 81, 85, 124, 186
classical physics 3, 17, 20, 47, 144
climate change 196
closed universe 15
collapse of the wave function 23, 28, 149
collective unconscious 69, 89, 115, 143, 146, 173
communism 191, 198
complementarity 22, 23, 43, 144, 183
consumerism 177, 178, 191, 192
Copenhagen Interpretation 149
cosmic microwave background radiation 11, 14, 100
cyanobacteria 104

D

dark energy 13, 14, 117, 182
dark matter 14, 100, 182
Darwin, Charles, 3, 106, 108, 130, 131, 132, 151, 207
Dawkins, Richard, 131, 202, 207
democracy 190
Descartes, Rene, 3, 142, 146, 187
d'Espagnat, Bernard, 125, 202, 207
devayonis (luminous bodies) 128
devotion 128, 161, 199
dharana 60, 152, 163
dharma 155
dhyana 60, 152, 163
diffraction pattern 26
DNA 87, 117, 130, 131, 133, 135, 140
dogma 85, 153, 178, 193, 196, 200
Dossey, Larry, 54, 90, 203, 204, 206, 208
dreams 70, 75, 124, 161, 185
dualism 7, 151, 152

E

Earth 4, 11, 27, 31, 33, 34, 37, 38, 42, 84, 90, 93, 97, 103, 104, 134, 135, 181
ecstasy 61, 63, 115, 116, 160, 173, 185, 201
EEG 74, 87
ego 60, 63, 106-109, 113, 123, 126, 149, 150, 157, 163, 165, 168-176, 199
Einstein, Albert, 1, 4, 11, 17, 18, 19, 20, 31, 32, 37, 43, 44, 100, 149, 150, 181, 208
EKG 53, 54
electromagnetic force 135
engrams 55
enlightenment 7, 60, 114, 167, 201
entanglement 29, 30, 44, 187, 188
entropy 39, 40
epiphenomenalism 3, 76, 153
evil 95, 108, 155, 157-159
evolution 3, 47, 92, 95, 104, 105, 106-108, 110, 112, 130-132, 135, 145, 151, 157, 168, 186
evolution of species 130-132, 151
extrasensory perception (ESP) 6, 49, 54, 69, 71, 78, 79, 81, 85, 86, 88-93, 116, 129, 153

F

fine-tuning 15, 135, 136, 137
Flatlander 36
flat universe 13, 14, 15
free will 39, 107, 109
fundamental factors 100, 102, 103, 107, 117, 118, 128, 183

G

Galileo 68, 93, 142
ganzfeld telepathy experiments 85
gauge bosons 136
general theory of relativity 32
ghosts 126, 127, 128

gilgul 67
Gisin, Nicolas, 20
Global Consciousness Project 84
gluon 137, 138
Golden Rule 123, 154, 199
good 70, 83, 95, 108, 115, 116, 122, 123, 154-158, 163, 169, 170, 175, 178, 186, 192, 197, 199
gravitational lensing 182
graviton 137
gravity 13, 32, 34, 41, 79, 101, 135, 136, 137, 149, 182, 183
guna 98
guru v, 60, 126, 152, 165, 173

H

hallucination 70, 99, 125, 127
Hammons, Ryan, 64, 65
happiness 62, 95, 109, 155, 156, 160, 161, 166, 170-172, 174, 175, 192, 197, 199, 201
Hawking, Stephen, 15, 134, 202, 207
Heisenberg uncertainty principle 139
Heisenberg, Werner, 22, 139, 147, 202, 208
hemophiliacs 131
Higgs boson 137, 138
Hinduism 60, 67, 115, 162
hippocampus 164
homeopathy 193
homunculus 56
Hubble, Edwin, 11, 13
Hubble's law 11
hydrogen 4, 5, 11, 17, 40, 50, 100-102, 135
hypnosis 52, 53, 55, 59, 89, 115, 124, 127
hypnotic regression 66, 124

I

idealism 2, 6
ideology 1, 6, 7, 45, 97, 113, 117, 121, 137, 150, 154-158, 177-179, 182-184, 190, 191, 193, 195-201
IIED (see intention imprinted electrical devices) 87
Indian rope trick 125
inflation 14, 15, 136
irreducibly complex 131, 132
ishta 165
ishta mantra 165
Islam 67, 151, 158

J

Jeans, James, 145, 146
Jesus 60, 67, 68, 73, 89, 174, 204
jihadism 158
jihadist 193
John the Baptist 67
Judaism 60, 67, 151
Jung, Carl, 115, 143, 146, 148, 171, 207, 209
Jupiter 103

K

karma 89, 95, 126, 155, 156, 169, 170, 172, 175
Kepler, Johannes, 143
kirtan 165, 199
Krishna 59, 62, 73, 203
Krishna, Gopi, 59, 62, 73, 203
kundalini 62, 119, 152, 165

L

Lao Tzu 60
Large Hadron Collider 137
law of conservation of energy 92
Law of Karma 122, 154, 191, 197
layers of mind 113, 114, 124, 129, 152, 171, 208
liberal education 190

INDEX

light 3-5, 11, 12, 14, 15, 17-20, 23-25, 27, 29, 31-34, 38, 41-43, 61, 62, 72, 75, 76, 79, 87, 88, 99-101, 107, 117, 127, 135, 149, 158, 181-183, 188
liila 110
Livestock and Climate Change (LCC) 196, 209
Long, Jeffery, 73, 205, 209
LSD 70, 75, 162
luminous factor 100, 101, 103, 107, 128

M

madhuvidya 175
manomaya kosha 114
mantra 152, 164, 165
marga 108
Margenau, Henry, 148, 149, 208
Mars 103
Martyn, Marty, 65
materialistic worldview 4, 5, 6, 44, 63, 78, 84, 92, 95, 151, 153, 177, 179, 181, 187, 188, 190, 191
memory 49, 54,-57, 61, 66, 72, 74, 75, 87, 114, 115, 133, 134, 170
Mercury 103
metaphysics 6, 93, 130, 142-144, 151, 153
meteorites 103
Michelson, Albert, 4, 17, 31
Milky Way 104
mindfulness meditation 163, 164
mind-stuff 7, 99, 105-107, 114, 115, 121, 122, 127
Mohammed 60
moksha 60
monism 6, 7, 144, 177
monistic 2, 6, 45, 67, 152
monistic idealism 2, 6
Moody, Raymond Jr., 66, 73, 75, 76, 203, 204, 205
morality 126, 154, 163, 199

Morley, Edward, 4, 17, 31
Moses 60
Mozart 68
mukti 60
muladhara chakra 118, 119
multicellular organisms 104, 105, 106, 117
mutative force (or principle) 99
Myers, F.W.H., 79
mystical experience 7, 49, 59, 60, 61, 116, 145, 162

N

nadis 118, 119
nationalism 190
natural selection 131, 132
naturopathy 193
near-death experience (NDE) 72-76
Nelson, Roger, 84
Neptune 103
neuron 56, 58, 92, 180
Newton, Sir Isaac, 3, 5, 17, 32, 33, 68, 122, 142, 153
nirvana 7, 60, 61
niyama 60, 154
nocebo 52

O

occult powers 126
Ockham's razor 137
ontology 48, 63, 94, 95, 108, 143, 144, 179, 181, 184, 187, 193, 200
open universe 13, 15
out-of-body experiences (OBE) 50, 70-72, 76, 185, 204

P

panentheism 151
pantheism 151
paradigm shift 1, 8, 177, 178
paranormal beings 126

parapsychology 78, 79, 85
Parnia, Sam, 71, 74, 204
Patanjali 60, 152, 154, 162, 163
Pauli, Wolfgang, 143, 144, 208
PEAR, Princeton Anomalies Research Laboratory 84, 86, 87
Penrose, Roger, 136, 202, 207
photoelectric effect 17, 149
photon 5, 20, 24-27, 29, 137, 138
photosynthesis 104, 117
placebo 49, 51, 52, 185
Planck, Max, 12, 17, 22, 43, 100, 143, 208
Planck's constant 22, 143
Planck spacecraft 100
plasma 100, 101, 102
Plato 44, 73, 146
polarization 19, 20, 190
Prakriti 98, 99, 105
prana 119, 163
pranayama 60, 163
prati-saincara 102, 104, 106
pratyahara 60, 163
prayer 54, 55, 162, 193
precognition 78, 81, 82, 85, 124, 186
presentiment 82, 85
probability 16, 18, 24, 28, 91, 139, 140, 149
proteins 133, 135
protoplanets 102
provincialism 190
psychic abilities 89, 124
psychic body 71, 117-121, 128, 165, 208
psychic phenomena 6, 49, 88, 93, 94
psychokinesis 78, 81, 83, 85, 124, 186
Purusha 98

Q

Qualifying Principle (see also Prakriti) 98, 99, 110, 119, 157, 173
quantum 1, 5-7, 12, 14, 17-30, 42-44, 57, 58, 79, 83, 92, 95, 100, 106, 124, 125, 137, 143, 145-150, 180, 183, 184, 186, 187
quantum physics 12, 24, 28, 143, 145, 147-150, 186, 187

R

Radin, Dean, 83, 205, 206, 208
rajoguna 98
Ramakrishna 164
reactive momenta 121-123, 126, 128, 155-157, 159, 170, 174, 175, 185
reciprocal apparitions 71
Rees, Sir Martin, 136, 207
reincarnation 6, 65-69, 89, 108, 124, 197
relativity theory 1, 5, 42, 43, 95, 100, 150
religious fundamentalism 193, 194
rishi 173
RNA 133

S

Sadashiva 1, 60, 162
sadhaka 172
sadhana 152
sage 162, 164, 173
sahasrara chakra 118
saincara 102, 104, 106
samadhi 7, 53, 60, 113, 152, 162
samskara (see also reactive momenta) 121, 122
Sanskrit 98, 105, 108, 109, 113, 115, 118, 119, 126, 157, 164, 165, 175, 200
Sarkar, P.R., 97, 98, 198
Satan 157
sattvaguna 98, 99
Saturn 103

Satyamurti 53, 54
Schrödinger, Erwin, 28, 146, 147, 148, 208
Schwartz, Gary, 88, 206
scientific epistemology 18, 44
seeker 169, 171-173, 176, 197
seer 173, 199, 201
self-realization 7, 60, 113, 154, 155, 161-163, 165, 167, 168, 171, 172
Sheldrake, Rupert, 118, 207
Shiva 1
siddhis 71, 89, 125
sin 151, 155-157, 191
socialism 198
soul 3, 61, 63, 67, 68, 109, 116, 128, 142, 156, 198, 199, 201
space-time 5, 12, 19, 23, 31-44, 78, 81, 99, 100, 117, 125, 134, 139, 149, 150, 179, 181-183, 187
spatial nonlocality 24, 179
special theory of relativity 4, 31, 32, 44
spirituality 3, 6, 7, 97, 132, 172, 193
spiritual worldview 2, 6, 7, 45, 69, 95, 97, 121, 125, 130, 141, 142, 147, 148, 153, 154, 155, 158, 177, 178, 181, 183, 184, 187, 188, 196, 197
SRI 88
Standard Model of particle physics 136
stars 13, 14, 99, 101, 102, 104, 134, 135, 145, 207
static force (or principle) 98-102, 113
St. Catherine of Siena 61
St. Francis of Assisi 52
stigmata 49, 52, 185
strong nuclear force 135, 137
subconscious 59, 74, 114, 116, 127, 171
suffering 52, 66, 74, 89, 95, 122, 124, 155, 156, 157, 159, 173, 174, 201
Sufism 7, 60, 67, 162
sun 33, 36, 40, 90, 93, 101, 103, 104, 112, 134, 145, 173
superconscious mind 115
supernova 33, 101, 181, 207
supersymmetry 137
svadhisthana chakra 118
swastika 115

T

tamoguna 98, 99
Tantra 7, 97, 162
Taoism 7, 67, 162
telepathy 78-81, 85, 124, 186
temporal nonlocality 24-26, 179
time travel 40, 42
tsunami 91
Tucker, Jim, 64, 65, 203

U

uncertainty principle 20, 22, 23, 124, 139, 147
United Nations Food and Agriculture Organization 196
unit mind 104-109, 114, 115, 118, 121-123, 155, 156, 184
unity of self-awareness 49
Upanishads 147
Uranus 103

V

Vedantism 7, 67
Venus 103
vidya 108, 109, 120, 156, 157, 159, 162, 165, 172
vishuddha chakra 118
vitalism 68
Vivekananda 162

W

wave function 7, 23, 24, 27, 28, 43, 57, 58, 83, 92, 93, 124, 129, 138-140, 149, 179, 180, 183
weak force 135
wormhole 42

Y

yama 60, 154
yoga 7, 60, 71, 72, 97, 109, 119, 152, 154, 162
Yoga Sutras 60, 154, 162

www.ingramcontent.com/pod-product-compliance
Lightning Source LLC
Chambersburg PA
CBHW021059080526
44587CB00010B/311